ELEMENTARY SCATTERING THEORY

Elementary Scattering Theory

For X-ray and neutron users

D.S. SIVIA

St John's College, Oxford

OXFORD
UNIVERSITY PRESS

OXFORD
UNIVERSITY PRESS

Great Clarendon Street, Oxford OX2 6DP

Oxford University Press is a department of the University of Oxford.
It furthers the University's objective of excellence in research, scholarship,
and education by publishing worldwide in

Oxford New York

Auckland Cape Town Dar es Salaam Hong Kong Karachi
Kuala Lumpur Madrid Melbourne Mexico City Nairobi
New Delhi Shanghai Taipei Toronto

With offices in

Argentina Austria Brazil Chile Czech Republic France Greece
Guatemala Hungary Italy Japan Poland Portugal Singapore
South Korea Switzerland Thailand Turkey Ukraine Vietnam

Oxford is a registered trade mark of Oxford University Press
in the UK and in certain other countries

Published in the United States
by Oxford University Press Inc., New York

© D.S. Sivia 2011

The moral rights of the author have been asserted
Database right Oxford University Press (maker)

First published 2011

British Library Cataloguing in Publication Data
Data available

Library of Congress Cataloging in Publication Data
Data available

Typeset by the author in LaTeX
Printed and bound by
CPI Group (UK) Ltd, Croydon, CR0 4YY

ISBN 978–0–19–922867–6 (Hbk.)
      978–0–19–922868–3 (Pbk.)

To Richard and Roger

Preface

The opportunities for doing experiments at synchrotron and neutron facilities have grown rapidly in recent years, and are set to continue into the foreseeable future. Powerful third generation X-ray sources, comprising the APS in Chicago, the ESRF in Grenoble and SPring-8 in Hyōgo, which became operational in the 1990s, are being joined by an increasing number of new regional facilities; DIAMOND and SOLEIL, for example, to name but two. Similarly, the availability of neutrons is being enhanced by the upgrade and expansion of established facilities, like the ILL in France and ISIS in England, and the commissioning of next generation sources, such as the SNS in the United States and J-PARC in Japan.

The early years of X-ray and neutron scattering were dominated by physicists, who traditionally had a strong background in the relevant theory and mathematics needed to understand the nature of the experiments; erudite texts, such as Lovesey (1986), could be consulted when details were required. The recent rise in the accessibility of synchrotrons and neutrons has been accompanied, if not driven, by a shift in emphasis from the study of hard condensed matter to the investigation of biological samples and soft matter. As a result, the demographics of the users at the respective facilities has been changing rapidly: an increasing proportion are biologists and chemists, for whom the scholarly physics-oriented works are generally too demanding to be helpful. This book aims to provide a gentler introduction to the subject, to cater for those who are less confident in their mathematical ability and physics background.

It could be argued that a knowledge of scattering theory is not necessary for being a user at a synchrotron or neutron facility these days, because the techniques are sufficiently well-developed to be treated as standard tools where the required data analysis is carried out with tried-and-tested computer programs. While there is truth in this viewpoint, a basic understanding of the underlying theory is always intellectually fulfilling and useful: it educates us about what can be inferred reliably from the measurements and, hence, influences which experiments are done. It is also important to remember that no analysis program is ever fool-proof or free from assumptions and approximations; consequently, it is advantageous to have some idea of how the data are related to the object of interest so that the numerical results can be checked against common sense.

Although mathematics cannot be avoided in a theoretical discussion, my aim has been to write a book that most scientists will find approachable. To this end, the first two chapters are devoted to providing a tutorial on background material that is implicitly assumed in other texts. Thereafter, the philosophy has been one of keeping things as simple as possible: clarity and understanding have been favoured over a comprehensive treatment, through the omission of topics and detail that entail too much complexity. Despite my best efforts, some students will still find this book quite daunting. If they can get something out of it, then I will consider the endeavour to have been worthwhile.

Finally, it is a great pleasure to acknowledge the help and encouragement that I received from a number of friends and colleagues. First and foremost, Doryen Bubeck, Steve Collins, Bill Hamilton, Jerry Mayers, Jeff Penfold, Dave Waymont and John Webster, who gave me useful feedback on many of the chapters. At various points, I've also picked the brains of Nikitas Gidopoulos, Stephen Lovesey, Roger Pynn and Jorge Quintanilla. Assistance and advice on individual topics and chapters was given by Peter Battle, Iain Campbell, Marilyn Hawley, Steve King, Kevin Knight, Stewart Parker, Owen Saxton and Mona Yethiraj. Many thanks go to my Editor, Sönke Adlung, for his patience; and to Emma Lonie, April Warman and other staff at the Oxford University Press. Last but not least, I am grateful to *Unilever R&D* for their generous sponsorship of this book, and to Ian Tucker for his help in securing it.

Oxford
July, 2010   D.S.S.

Contents

Part I

Some preliminaries

Studying matter at the atomic and molecular level

1

The world around us is varied and complex but, as scientists, we hope to make sense of it by trying to understand the nature of its underlying constituents. The notion that the universe is made up of basic building blocks goes back to antiquity, with the word 'atom' itself coming from the Greek *atomo* meaning 'indivisible'. Although this idea was put on a firm footing by chemists such as Dalton and Mendeleev in the nineteenth century, a theoretical picture of atoms did not emerge until the work of physicists like Thompson, Rutherford and Chadwick in the early part of the twentieth century. While subsequent experiments in high energy physics have revealed a rich subatomic structure of quarks and so on, our interest here is with the study of condensed matter (liquids and solids) at the atomic and molecular level.

1.1 Length scales and logarithmic axes

The SI, or *Système International*, unit of length is a *metre* and has a standard abbreviation of 'm'. It is a sensible measure for everyday objects, such as people, because their size tends to be of this order. The breadth of a human hair, which has traditionally been used as a metaphor for something extremely narrow, is about ten to twenty thousand times smaller than a metre. Rather than using the fractional notation of $1/20000$ m or 0.00005 m, which becomes increasingly cumbersome as the numbers involved get very small (or large), we tend to write it as 5×10^{-5} m or 0.05 mm or $50 \, \mu$m, where the abbreviated prefixes of 'm' (for milli) and 'μ' (micro) represent a multiplication by 10^{-3} and 10^{-6} respectively.

The most common SI prefixes are listed in Fig. 1.1, which shows the corresponding magnitude factors drawn on a *logarithmic* scale. Whereas uniform increments on a *linear* (ordinary) axis denote the addition of a constant at each step, they indicate a multiplication by a given scale-factor on a logarithmic one; this difference is illustrated in Fig. 1.2. In the case of Fig. 1.1, each (major) tick mark corresponds to a thousand-fold increase in magnitude over its predecessor on going up the axis. Everyday examples of logarithms include musical scales, where successive *octaves* double the frequency of a note, the *Richter* scale for indicating the severity of an earthquake,

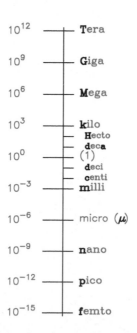

Fig. 1.1 The standard SI prefixes.

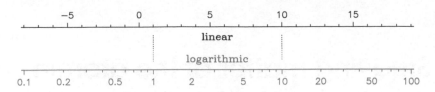

Fig. 1.2 The difference between linear and logarithmic axes.

the *decibel* scale for reporting the power gain of a hi-fi amplifier, the decay of a radioactive sample as specified by its *half-life*, the growth of a bacterial colony as defined by its 'doubling-time', the pH scale for acidity and so on.

From Sivia and Rawlings (1999), *Foundations of Science Mathematics*, Oxford Chemistry Primers Series, **77**.

$a^1 = a$
$a^2 = a \times a$
$a^3 = a \times a \times a$
$a^4 = a \times a \times a \times a$

Powers, roots and logarithms

If a number, say a, is multiplied by itself, then we can write it as a^2 and call it a-squared. A triple product can be written as a^3 and is called a-cubed. In general, an N-times self-product is denoted by a^N where the superscript, or *index*, N is referred to as the *power* of a.

Although the meaning of 'a to the power of N' is obvious when N is a positive integer (i.e. $1, 2, 3, \ldots$), what happens when it's zero or negative? This question is easily answered once we notice that the procedure for going from a power of N to $N-1$ involves a division by a. Thus if a^0 is a^1 divided by a, then a to the power of nought must be one (or *unity*); similarly, if a^{-1} is a^0 divided by a, then it must be equal to one over a; and, in general, a^{-N} is equivalent to the *reciprocal* of a^N. Thus, we have

$$a^0 = 1 \quad \text{and} \quad a^{-N} = \frac{1}{a^N} \, . \tag{1.1}$$

The basic definition of powers leads immediately to the formula for adding the indices M and N when the numbers a^M and a^N are multiplied:

$$a^M a^N = a^{M+N} \, . \tag{1.2}$$

$a^{1/2} \, a^{1/2} = a$

$a^{1/3} \, a^{1/3} \, a^{1/3} = a$

While our discussion has so far focused only on the case of integer powers, suppose that we were to legislate that eqn (1.2) held for all values of M and N. Then, we would be led to the interpretation of fractional powers as *roots*. To see this, consider the case when $M=N=1/2$; it follows from eqn (1.2) that a to the power of a half must be equal to the square root of a. Extending the argument slightly, if a to the power of a third is multiplied by itself three times then we obtain a; therefore, $a^{1/3}$ must be equal to the cube root of a. In general, the pth root of a is given by

$$a^{1/p} = \sqrt[p]{a} \tag{1.3}$$

where p is an integer. One final result on powers that we should mention is

$$\left(a^M\right)^N = a^{MN} \, , \tag{1.4}$$

which can at least be verified readily for integer values of M and N.

An alternative method of describing a number as a 'power of something' is to use *logarithms*. That is to say, if y is written as a to the power of x then x is the logarithm of y to the *base* a:

$$y = a^x \qquad \Longleftrightarrow \qquad x = \log_a(y) \qquad (1.5)$$

where the double-headed arrow indicates an equivalence, so that the expression on the left implies the one on the right and vice versa. Since we talk in powers of ten in everyday conversations (e.g. hundreds, thousands, millions), the use of $a = 10$ is most common; this gives rise to the name *common* logarithm for $\log_{10}$, often abbreviated to just log (but this can be ambiguous). Other bases that are encountered frequently are 2 and 'e' $(2.718\ldots)$; $\log_e$, or ln, is called the *natural* logarithm.

By combining the definition of the logarithm in eqn (1.5) with the rule of eqn (1.2), it can be shown that the 'log of a product is equal to the sum of the logs' (to any base); and, in conjunction with eqn (1.1), that the 'log of a quotient is equal to the difference of the logs':

$$\log(AB) = \log(A) + \log(B) \qquad \text{and} \qquad \log(A/B) = \log(A) - \log(B). \quad (1.6)$$

Similarly, eqn (1.5) allows us to rewrite eqn (1.4) in terms of the log of a power and to derive a formula for changing the base of a log (from a to b)

$$\log(A^\beta) = \beta \log(A) \qquad \text{and} \qquad \log_b(A) = \log_a(A) \times \log_b(a). \quad (1.7)$$

Thus $\ln(x) = 2.3026 \log_{10}(x)$, where the numerical prefactor is $\ln(10)$ to four decimal places, and so on.

$$\log_{10}(10^0) = 0$$
$$\log_{10}(10^1) = 1$$
$$\log_{10}(10^2) = 2$$

$$\ln(e^0) = 0$$
$$\ln(e^1) = 1$$
$$\ln(e^2) = 2$$

$$\exp(x) = e^x$$

While hair may be extremely narrow, it is made of cells that are about ten times smaller; at around $10\,\mu$m, they are not visible (or differentiable) to the naked eye. The simplest single-celled organisms, or *prokaryotes* such as bacteria, are typically a tenth of the size of the average human cell; that is to say, about a micrometre (or a *micron*). Viruses, which are little more than packets of genetic material surrounded by a coating of proteins, are a magnitude smaller at around $0.1\,\mu$m or 100 nm. Whether it be viruses or organelles within cells, such as *ribosomes* where the genetic code is translated into proteins, they are essentially assemblies of large molecules. Although these macro-molecules can be very long if stretched out like a string, with the human genome being 1.8 m for example, they tend to be twisted, folded, packed and clumped into units of order 10 nm. The nucleic acids (DNA and RNA), proteins and complex carbohydrates are themselves made up of building blocks of smaller molecules (the nucelotide bases, amino acids and simple sugars), with dimensions of a nanometre.

The shortest length scale of interest to us pertains to inter-atomic distances, which are of order 0.1 nm or 100 pm. We often refer to it by a special unit named after the physicist Ångström (1814–1874):

$$1\,\text{Å} = 10^{-10}\,\text{m} = 0.1\,\text{nm} = 100\,\text{pm}.$$

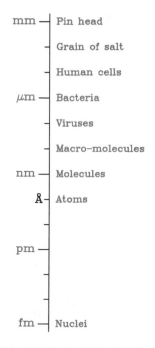

Fig. 1.3 Physical length scales.

The nuclei of atoms, and the neutrons and protons of which they consist, are ten to a hundred-thousand times smaller than an angstrom: around 10^{-15} m or a few femtometres. The main point to note here is that the nuclei are point-like when compared with the dimensions of atoms.

1.2 Resolution, magnification and microscopy

Atoms cannot be seen with the naked eye because they're too small, but what limits our ability to see objects below a certain size? It's due to the wave-nature of light, which bends around edges, such as the sides of the pupils in our eyes, and causes us to see a slightly blurred view of the world. This effect can be noticed with large televisions where the individual pixels can be distinguished when standing close to the screen but not on moving further away; they appear to run into each other and merge to give a seamless picture. This observation tells us that the limit on seeing detail depends on the angular size of the object rather than its intrinsic dimensions. While the diameter of the sun is 400 times larger than that of the moon, for example, it has the same projection of about $0.5°$ on the sky because it's also 400 times further away.

For any optical device or imaging system, the smallest angular scale, $\delta\theta$, that can be resolved is given by

$$\boxed{\delta\theta \approx \frac{\lambda}{D}} \,, \tag{1.8}$$

From Sivia and Rawlings (1999), *Foundations of Science Mathematics*, Oxford Chemistry Primers Series, **77**.

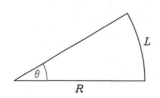

$$1 \text{ rad} = \frac{180°}{\pi} \approx 57.3°$$

Angles and circular measure

An angle is a measure of rotation, or turn, and is usually specified in *degrees*. There are $360°$ in a complete twist, where one ends up facing the same way as at the beginning; $90°$ therefore represents a right-angle, $180°$ an about-face, and so on. Despite our familiarity with degrees, a circular measure that is often more useful in a mathematical context is a *radian*. It is a dimensionless quantity defined as follows: if a radius of length R is spun through an angle θ and generates an arc of length L, then

$$\theta = \frac{L}{R} \,. \tag{1.9}$$

Since the circumference of a circle is $2\pi R$, $360° = 2\pi$ radians; a right-angle is $\pi/2$ and, in general, an angle in degrees can be converted into one in radians by multiplying it by $\pi/180$. Unless degrees are mentioned explicitly, it is best to assume that all angles are implicitly given in radians (especially if they contain factors of π).

where λ is the wavelength of the light, D is the size of the aperture through which the observation is made and $\delta\theta$ is in radians. Taking the wavelength of light to be 600 nm and the diameter of the pupil as 2 mm, eqn (1.8) yields a resolution limit of

$$\delta\theta \approx \frac{600 \times 10^{-9}}{2 \times 10^{-3}} = 300 \times 10^{-6} \text{ rad} \approx 0.017°,$$

or about one minute of arc, for the naked eye. Given that most people have difficulty in focusing on objects that are closer than 20 cm from their eyes, we can use the angular definition of eqn (1.9) to relate this $\delta\theta$ to a minimum resolvable size of around 60 μm. Since this is roughly the breadth of a human hair, there is some wisdom behind the saying 'as narrow as a hair's breadth'!

We should emphasize that the above analysis does not imply that things that are much smaller that 60 μm cannot be detected with the naked eye. If they occur in isolation, and are sufficiently bright or well-lit, they should be visible. If two are closer together than 60 μm, however, we will not be able to distinguish them as separate entities; this loss of detail is the significance of resolution. The formula of eqn (1.8) is based purely on the *diffraction* of light through an aperture and represents the best limit achievable in theory; the resolution attained in practice can be worse for a variety of reasons. Poor lighting conditions lead to a larger $\delta\theta$, for example, even though the pupils then expand to $D \approx 8$ mm. Likewise, a failure to wear spectacles, if required, degrades the resolution.

Things that are too small to be seen easily with the naked eye can be made more accessible, of course, with the aid of a magnifying glass. Although references to magnifiers and 'burning glasses' are found in the writings of Pliny the Elder (23–79 AD), and Egyptian artefacts from 640 BC include rock crystals in the shape of *convex* lenses, their use was very limited until the invention of spectacles towards the end of the thirteenth century. The greatest magnification from a lens is attained when the object of interest is placed a distance d from it that is comparable to its *focal length* f. Since a lens also constitutes an aperture whose diameter cannot be larger than $2f$ (which would result from a spherical lens!), it yields a diffraction limit on the smallest resolvable size, δL, of around half the wavelength of light:

$$\delta L \approx \frac{\lambda}{D} \times d > \frac{\lambda}{2f} \times f.$$

While microscopes are more complex than simple magnifying glasses, the limit given by this elementary argument still holds: the shortest length scale that can be resolved by the best optical instruments is of order 0.2 μm (with $\lambda \approx 400$ nm for blue light).

The above analysis tells us that, to see finer detail, we have to use light of a shorter wavelength. In particular, to achieve atomic resolution, we must move beyond the visible and ultraviolet regions of the *electromagnetic spectrum* (Fig. 1.4) to X-rays. There is a serious

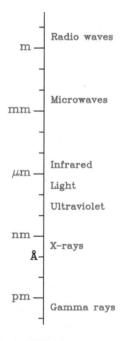

Fig. 1.4 The electromagnetic spectrum.

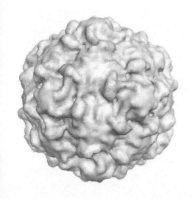

Fig. 1.5 A poliovirus, 30 nm across, as seen by a transmission electron microscope. (Bubeck *et al.*, 2005)

All these fifty years of conscious brooding have brought me no nearer to the answer to the question, 'What are light quanta?' Nowadays every Tom, Dick and Harry thinks he knows it, but he is mistaken.

(**Albert Einstein, 1954**, writing to his friend Michael Besso.)

problem, however, in that it's very difficult to produce good lenses for light of such short wavelengths; hence, X-ray microscopy is not a practical way forward.

A solution to the quandary of having very short wavelength radiation that can be focused easily is provided by the *electron*: as a charged particle, its trajectory can be controlled with both electric and magnetic fields; thanks to *quantum mechanics*, it also exhibits wave-like behaviour. Electrons are readily produced by a hot wire and then accelerated to the desired speed by a potential difference of V volts; the only substantial requirement is that of a vacuum, to avoid energy losses through collisions with air molecules. The associated wavelength is given by eqn (1.13) as

$$\lambda = \frac{h}{\sqrt{2\,m_e\,eV}} = \frac{1.23 \times 10^{-9}}{\sqrt{V}}\ \mathrm{m}\,, \tag{1.10}$$

where the momentum p of the electron has been related to its mass $m_e = 9.11 \times 10^{-31}$ kg and charge $e = 1.60 \times 10^{-19}$ C through the conservation of energy (electrostatic to kinetic) in the *classical* limit:

$$eV = \frac{p^2}{2\,m_e}\,. \tag{1.11}$$

Wave–particle duality and quantum mechanics

We are familiar with the difference between the interactions of discrete objects compared with those of waves. Two colliding balls, for example, either bounce off each other or stick together; ripples on a pond, on the other hand, simply pass through each other unaffected but combine to give an interference pattern while overlapped. This everyday experience of the distinct characteristics of waves and particles does not seem to hold at the microscopic scale: objects that behave like particles in one situation display phenomena associated with waves in another. Although the nature of light has been a source of much debate since the 17th century, with Hooke supporting Huygen's wave theory while Newton regarded it as a tiny stream of particles ('corpuscles') travelling at high speed in straight lines, the inherently dual nature of matter at the smallest scales did not become generally apparent until the late 19th and early 20th centuries. It led to the birth of quantum mechanics.

The wave–particle duality of nature is captured at the most elementary level by Planck's hypothesis (1901) that an oscillation of frequency ν is associated with a packet, or 'quantum', of energy E given by

$$E = h\nu, \tag{1.12}$$

where $h = 6.626 \times 10^{-34}$ J s is called Planck's constant, and de Broglie's proposal (1923) that a particle with momentum p is associated with a wavelength λ given by

$$\lambda = \frac{h}{p}\,. \tag{1.13}$$

Only a hundred volts are needed, therefore, to accelerate electrons to wavelengths comparable to atomic length scales. More energetic electrons are used in practice ($\sim 200\,\text{keV}$), to enable them to pass through thicker samples and, perhaps surprisingly, to reduce the radiation damage. The latter imposes a serious limitation on the image resolution attainable with biological materials, because they degrade rapidly with exposure, and it is lower than that for inorganic samples by an order of magnitude. An image of a 30 nm poliovirus, with the best resolution currently achievable ($\sim 10\,\text{Å}$), is given in Fig. 1.5; and an inorganic example, where individual atoms can just be differentiated, is shown in Fig. 1.6.

The *electron microscopy* considered above is the analogue of its optical counterpart, where the illuminating radiation passes through a suitably thin sample. An alternative to this 'transmission' mode of operation (TEM) is a 'scanning' version (SEM), where a highly focused beam of electrons is directed at successive small areas of the surface and a map constructed from the returned signal (principally of secondary emissions). Although the resolution achieved with SEMs is about an order of magnitude poorer than that of TEMs, the resultant images have a good depth of focus and are much easier to interpret.

An even more direct approach to the study of surfaces is provided by *scanning probe microscopes* (SPMs), where an extremely sharp probe is raster-scanned in close proximity to the sample and topographic maps constructed from a variety of interaction signals induced in the tip. Scanning tunneling microscopes (STM) were the first of this type of instrument, being developed in 1981, but many different kinds of atomic force microscopy (AFM) now exist. SPMs typically achieve a resolution in the nm to Å range, can be operated in a wide range of environments and temperatures, and work equally well for metals and soft materials. An example of an STM image is given in Fig. 1.7, which shows a rectangular array of atoms on a (clean) surface of silicon.

Fig. 1.6 A transmission electron micrograph of cadmium sulphide. (Courtesy of Dr. J. H. Warner, Department of Materials, Oxford.)

Fig. 1.7 An STM image of atoms on the surface of silicon, $3.84\,\text{Å}$ and $7.68\,\text{Å}$ apart, taken under ultra-high vacuum conditions. (Courtesy of Dr. M. E. Hawley, Los Alamos National Laboratory.)

1.3 Structure, dynamics and spectroscopy

So far we have been thinking about looking at matter in the literal sense of an image. As in the case of an ordinary photograph, what appears like a frozen snap-shot from a potentially moving scene is actually a time-averaged view of the world. If the inherent motion is small compared with the resolution of the imaging system, or the equivalent of the camera's shutter-speed is sufficiently fast, the picture will be sharp; otherwise, it will be a blurred rendition of the object of interest. In either case, we can gain additional information on the nature and behaviour of matter at the atomic and molecular level by probing the *dynamics* on these length scales. This is the realm of *spectroscopy*.

The earliest form of spectroscopy consisted of the production of the colours of the rainbow when sunlight passed through a prism. While this had been known since antiquity, it was Newton (in 1666) who first realized the significance of this spectrum in terms of what it told us about the nature of light. With improving instrumentation, a large set of narrow dark lines became discernible in this spectrum by the early nineteenth century. Although Fraunhofer never understood the origin of these spectral lines that now bear his name, Kirchhoff subsequently discovered that their patterns were a unique signature of the chemical composition of the light source. Indeed, when the dark 'absorption' lines from the spectrum of sunlight were compared with the corresponding bright 'emission' ones obtained on placing different atoms and molecules in a (Bunsen) flame, an anomaly was found that led to the prediction (by Lockyer, in 1870) and later confirmation (by Ramsay, in 1895) of helium! It's named after the Greek sun god, *Helios*.

A study of the systematic patterns in the wavelengths of the spectral lines from elements such as hydrogen, by Balmer and others in the late 19th century, led to Bohr's quantum model of the atom in 1913. It shared Rutherford's picture of a compact positive core surrounded by orbiting negatively charged particles, much as planets going around the sun, but differed through its additional proposal that the *angular momentum* (and, hence, energy) of the electrons could only take certain well-defined values. The spectral lines then arose from transitions between the allowed energy levels (Fig. 1.8), with absorption resulting in an increase in the electron's energy and an emission occurring due to a corresponding decrease. Although this simple model was able to explain the spectra of hydrogen-like atoms (nuclei with a single *valence* electron), it failed to generalize to more complex cases. Nevertheless, it serves as a useful historical example of how structural information can be gleaned from spectroscopic measurements.

By the early 1800s, Herchel and Ritter had demonstrated that the visible spectrum was only a small portion of the light coming from the sun. Subsequently, absorption and emission lines were also found at ultraviolet (UV) and infrared (IR) wavelengths; what can be learnt from them? The higher frequency, or shorter wavelength, UV spectra give additional information on the energy levels of the valence electrons whereas the longer wavelength, or lower frequency, IR spectra tell us about molecular vibrations. Since the latter relate to the stretching and bending of bonds between atoms, IR spectroscopy gives an indication of the connectivity in molecules. Not only is it helpful in identifying the presence of particular bonds, but clues about their environments can also be obtained: the stretching frequency of the C=O carbonyl group, for example, varies slightly depending on whether it occurs in an aldehyde, a ketone, a carboxylic acid, an ester or an amide. Thus, structure and dynamics are intimately linked.

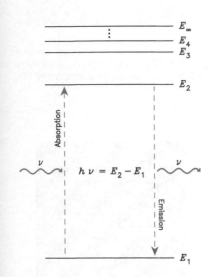

Fig. 1.8 A transition between two energy levels, E_1 and E_2, and the corresponding spectral line of frequency ν.

A simple example of quantization

If we pluck a string on a guitar, it vibrates and produces a note. What determines its frequency? Well, it's governed by the length, thickness and tension of the string; and, to a lesser extent, the nature of the plucking. The principal variable at the musician's disposal during a performance, therefore, is the length L, which can be changed by pressing a finger down on the relevant fret.

The simplest vibration of the string consists of an up-and-down motion of a *sine*-shaped displacement which is constrained to be immobile at the ends; the $n=1$ case in Fig. 1.9 shows a snap-shot. This is called the *fundamental harmonic*, and has a wavelength $\lambda=2L$. Higher-order (sinusoidal) harmonics, which have shorter wavelengths, can also be excited but must fulfill the condition that

$$L = n\lambda/2\,, \tag{1.14}$$

where $n = 1, 2, 3, \ldots$, to satisfy the immobility constraint at the two ends. This leads to the quantized normal modes shown in Fig. 1.9.

The same mathematics arises when considering the case of a particle of mass m trapped in a one-dimensional potential well of length L from a quantum mechanics point of view. The allowed de Broglie wavelengths of eqn (1.14) are then related to momenta $p_n = h/\lambda_n$, where $\lambda_n = 2L/n$, through eqn (1.13) and lead to the quantization of the associated kinetic energy $E_n = p_n^2/2m$:

$$E_n = \frac{n^2 h^2}{8mL^2}\,. \tag{1.15}$$

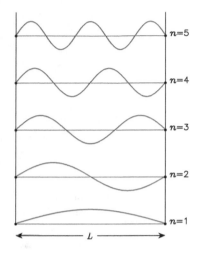

Fig. 1.9 The normal modes for a string of length L, fixed at both ends.

Absorption and emission spectroscopy can be extended even further from the visible spectrum, to microwave and X-ray wavelengths. While X-rays are related to transitions involving the deep electronic states of atoms, microwaves are associated with the excitations of molecular rotations (the basis of heat transfer in microwave ovens). There are many other forms of spectroscopy, most notably Raman and nuclear magnetic resonance (NMR), which provide information that is complementary to IR and microwave spectroscopy. Neutron scattering also has a role to play, albeit a niche one, as will be discussed in Chapter 9.

1.4 Atomic building blocks and interactions

We noted earlier that X-ray microscopes were not feasible because good enough lenses were not available for light of such short wavelengths. This lack of imaging hardware does not prevent us from examining how a beam of X-rays is scattered by a sample however, but what can we learn from such experiments? This question is at the heart of this book. To address it, we must first remind ourselves of the basic structure of atoms and their interactions.

1.4.1 The basic structure of the atom

In terms of the material in this book, and science in general, atoms can be viewed as being composed of three constituents: electrons, protons and neutrons. Although protons and neutrons are no longer considered to be fundamental, in the sense that they are made up of *quarks*, such subatomic subtleties are important only for particle physicists and cosmologists. So what are the basic properties of electrons, protons and neutrons, and how are they put together to form atoms?

As first suggested by Rutherford in 1911, a small and very dense nucleus lies at the heart of an atom. Its radius is ten to a hundred thousand times smaller than that of the atom (several femtometres rather than an angstrom), but it contains nearly all the mass. The nucleus is a conglomeration of positively charged protons and neutral neutrons; both have a similar mass, around 2000 times that of an electron. A 'diffuse cloud' of negatively charged electrons surrounds the nucleus, at a large distance from it, and gives the atom its overall size.

The element species is determined by the *atomic number*, which is the number of protons in the nucleus. This is also the number of electrons in a neutral atom. A loss or gain of orbital electrons yields an atom having a net positive or negative charge, respectively, and the resultant entity is called an *ion*. Neutrons play a role in the stability of the nucleus, and contribute about half the mass of the atom: the *atomic mass* is proportional to the total number of protons and neutrons. Atoms with the same number of protons but different numbers of neutrons are called the *isotopes* of an element, and are usually distinguished by specifying their atomic mass; for example, ^{235}U and ^{238}U for uranium. The isotopes of the simplest element (with just one proton) have been given special names: ^{1}H, which is the most abundant and lightest form (having no neutrons), is plain hydrogen; ^{2}H, which is twice as heavy (with one neutron), is called deuterium; ^{3}H, the heaviest 'stable' variant (with two neutrons), is tritium.

As well as the familiar properties of mass and charge, our three building blocks of nature have the additional quantum attribute of *spin*; this is difficult to visualize physically because it has no classi-

$^{\mu}_{z}X^{q}$ is the ionic isotope of element X, defined by its atomic number z, of mass μ and charge q.

Table 1.1 Attributes of the three principal constituents of atoms.

Particle	Charge ($\times 10^{-19}$ C)	Mass ($\times 10^{-27}$ kg)	Spin
Electron	-1.602177	0.000911	$\frac{1}{2}$
Proton	$+1.602177$	1.672622	$\frac{1}{2}$
Neutron	0	1.674927	$\frac{1}{2}$

cal analogue. Electrons, protons and neutrons all have a spin of $1/2$. Particles with half-integral spin ($1/2$, $3/2$, $5/2$, ...), such as these, are called *fermions*; only one can occupy any given quantum state. There is no occupational restriction for particles with integer spin ($0, 1, 2, \ldots$), which are known as *bosons*.

Although we no longer picture atoms as miniature solar systems with electrons going around the nucleus, quantum mechanics does yield a distinct set of allowed spatial distributions for the electrons called *orbitals*. They adopt the lowest energy states first but, being fermions, are forced to take up residence in ever higher ones as their number increases. The electrons in the outermost orbitals form the point of contact between atoms and determine how they combine to form molecules; this is the basis of chemistry.

Fig. 1.10 One of the d-shell electron orbitals pertinent to transition metals.

1.4.2 The fundamental forces of nature

The matter that makes up the world around us interacts through four fundamental forces: gravity, electromagnetism and the *weak* and *strong* nuclear forces. The first two act on everyday and astronomical objects, and have a strength that falls off with distance r according to the familiar $1/r^2$ inverse-square law. Gravity is by far the weaker of these interactions, as can be ascertained by comparing the magnitudes of the electrostatic and gravitational forces on a pair of isolated protons:

$$\frac{F_{\text{elec}}}{F_{\text{grav}}} = \frac{e^2/(4\pi\epsilon_{\text{o}} r^2)}{\mathrm{G}\, m_{\text{p}}^2/r^2} = \frac{e^2}{4\pi\epsilon_{\text{o}}\mathrm{G}\, m_{\text{p}}^2} \approx 10^{36},$$

where e and m_{p} are the charge and mass of the proton, respectively, G is Newton's universal constant of gravitation and ϵ_{o} is the permittivity of free space. Given the overwhelming strength of the electrostatic force, it can seem surprising that the motions of the stars, planets and galaxies are governed entirely by gravity. The reason is, of course, that charges tend to cancel out on the large scale, to leave essentially neutral bodies, whereas mass only adds up, to yield a cumulatively dominant attraction.

$$\mathrm{G} = 6.674 \times 10^{-11}\ \mathrm{N\, m^2\, kg^{-2}}$$
$$\epsilon_{\text{o}} = 8.854 \times 10^{-12}\ \mathrm{F\, m^{-1}}$$

Although gravity reigns supreme on a cosmological scale, electromagnetism is the principal driving force behind the chemistry of life and the physics of everyday objects. It controls how charged objects interact with each other and with electric and magnetic fields. At the nuclear level, however, it cannot explain the existence of atoms, because a conglomeration of protons should blow itself apart due to the repulsive electrostatic force between charges of the same sign. This must be overcome by an even stronger attractive force which is effective on length scales of femtometres; it is known as the 'strong nuclear force'.

The neutrons in the nuclei of some atoms spontaneously disintegrate into an electron, which is ejected as β-radiation, a proton and an antineutrino ($\bar{\nu}$). Neutrinos are ghostly particles, with no charge

$$n \longrightarrow p^+ + e^- + \bar{\nu}$$

Table 1.2 Attributes of the four fundamental forces.

Force	Strength (relative)	Range (m)	Mediator	Charge (e)	Mass (kg)	Spin
Gravity	1	∞	Graviton	0	0	2
Electromagnetic	10^{36}	∞	Photon	0	0	1
Weak	10^{25}	10^{-15}	Z^0, $W^\pm$	$0, \pm 1$	10^{-25}	1
Strong	10^{38}	10^{-15}	Gluons	0	0	1

$$_z^\mu X \xrightarrow{\beta} {}_{z+1}^{\mu}Y$$

and almost no mass (just spin), which are not of much concern to us, but the transmutation of a proton into a neutron increases the atomic number by one, thereby changing the identity of the atom, while leaving the mass virtually unchanged. This type of decay is controlled by a far less powerful short-range force called the 'weak nuclear force'.

In the nineteenth century, the experimental and theoretical work of Faraday and Maxwell showed that electricity and magnetism were intimately linked: a moving charge generated a magnetic field and a varying magnetic field induced current to flow in an electrical conductor. As such, electricity and magnetism were recognized as just being different manifestations of the same underlying phenomenon: electromagnetism. It has been a central goal of physics, since the late twentieth century, to understand the four forces discussed above in a similarly unified formalism. While some progress has been made, with the *electroweak* theory of Glashow, Salam and Weinberg in the 1960s combining electromagnetism and the weak interaction into a single force under suitably high energy conditions, a lot is still based on speculative ideas. Within the framework of these 'grand unified theories', or GUTs, force is seen as being mediated between the fermions that make up the building blocks of matter by 'virtual' bosons; these are particles (with integer spin) which come in and out of existence fleetingly, but in line with Heisenberg's *uncertainty principle* in quantum mechanics.

1.4.3 Probing matter by scattering particles

Having reviewed the basic structure of atoms and the fundamental forces, we can start thinking about how a beam of particles will interact with a sample when being scattered. The principal mechanism will depend on whether the incoming probe has a charge, if it has spin, the incident energy and so on. An electron, for example, will be electrostatically repelled by the orbital electrons of the atoms, thereby preventing it from penetrating the sample by more than a small fraction of a micrometre; as such, electron scattering is useful only for studying the surface layers of materials.

An X-ray photon may not have a charge but, as an electromagnetic wave, it does consist of oscillating electric and magnetic fields. The charge of the orbital electrons interacts with these fluctuating fields, but not as strongly as with a particle having a fixed charge. X-rays penetrate matter more easily than electrons, therefore, and are able to pass through a few millimetres of aluminium.

Neutrons are neutral particles that are unaffected by the charge of the orbital electrons. They interact with the nucleus of the atom via the strong nuclear force. Although inherently powerful, it acts only over very short distances which results in weak scattering. This offers the advantage of deep penetration (e.g. centimetres of aluminium), which enables the bulk properties of a material to be probed, but has the drawback of a poor scattering signal when a large sample is not available. If some of the atoms in the material under study are magnetic, by virtue of having unpaired orbital electrons, then the incident neutrons can be scattered by them through a spin-based dipole–dipole interaction. This mechanism can also operate in electron and X-ray scattering, but tends to be negligible compared to those mentioned above; with neutrons, however, nuclear and magnetic scattering occurs at a similar rate. This has made neutrons the dominant probe for studying magnetism at the atomic level, but the shear X-ray intensity of the new synchrotron sources is starting to make them a viable alternative.

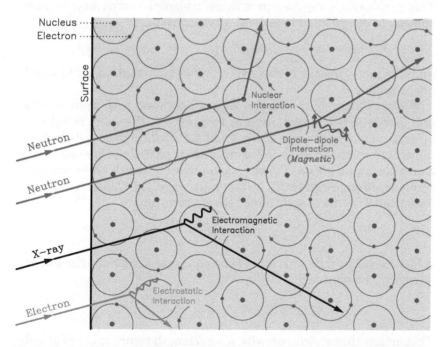

Fig. 1.11 A schematic illustration, following Pynn (1990), of the principal interactions by which electrons, X-rays and neutrons are scattered by condensed matter.

1.5 Energy, length and temperature scales

We began this introductory chapter by discussing length scales. Let's conclude by linking them with some corresponding measures of energy and temperature that are in common usage.

The central connection between length and energy comes from the original postulates of Planck and de Broglie, given in eqns (1.12) and (1.13), that were put forward to explain quantum phenomena. Starting with the former, a photon of frequency ν is associated with energy E through $E = h\nu$ where h is Planck's constant. Hence, X-rays can be characterized equally well by their energy as by their frequency or, through

$$\boxed{\nu = \frac{c}{\lambda}} \qquad (1.16)$$

$c = 2.998 \times 10^8 \, \text{ms}^{-1}$

where c is the speed of light, by their wavelength λ. Since the energies involved are tiny in the SI unit of joules, they are often stated in terms of the equivalent kinetic energy that an electron (with charge e) would gain when accelerated through a potential difference of V volts; 1 eV, or an electron-volt, is therefore about 1.6×10^{-19} J. These expressions for the energy yield the relationships

$$E = eV = h\nu = \frac{hc}{\lambda} . \qquad (1.17)$$

This enables an X-ray photon with a wavelength comparable to atomic spacing to be specified in several alternative ways:

$$\lambda = 1 \, \text{Å} \iff \nu = 3 \times 10^6 \, \text{THz} \iff E = 12.4 \, \text{keV} ,$$

and it is just a matter of context, or subject convention, as to which one is preferred.

Whereas it's easy to think of X-rays as waves, with their electric and magnetic fields oscillating at frequency ν, the same is not true of neutrons: they are one of the building blocks of atoms. It is more natural to consider them in terms of their particle properties, such as mass m_n and speed v. A connection with a length scale can then be made through eqn (1.13) by using $p = m_n v$, the classical mechanics formula for the magnitude of the momentum:

$$\lambda = \frac{h}{m_n v} .$$

A link to energy follows readily from equating E with the the kinetic energy of the neutron:

$$E = \tfrac{1}{2} m_n v^2 = \frac{h^2}{2 \, m_n \lambda^2} . \qquad (1.18)$$

This means that a neutron with a wavelength comparable to atomic distances can be specified equally well by

$$\lambda = 1 \, \text{Å} \quad \text{or} \quad v = 3.96 \, \text{kms}^{-1} \quad \text{or} \quad E = 81.8 \, \text{meV} .$$

We should note that chemists, especially spectroscopists, tend to use an alternative measure of energy called 'wavenumbers',

$$\text{Energy equivalent in wavenumbers} = \frac{E}{hc}, \qquad (1.19)$$

which corresponds to the reciprocal of the wavelength of a photon having energy E. The standard units are cm^{-1}, and are interpreted as being the number of wavelengths of light that would occupy a distance of $1\,\text{cm}$.

$1\,\text{meV} \equiv 8.066\,\text{cm}^{-1}$

Another useful measure of energy that is appropriate for particles such as neutrons is the temperature T: their average speed $\langle v \rangle$ is higher when they are in *thermal equilibrium* with something hot than when they are cold. In fact, an important result from thermodynamics, called the *equipartition of energy*, shows that each 'degree of freedom' (or movement) is associated with a mean energy of $\frac{1}{2}k_{\text{B}}T$, where temperature is measured in degrees Kelvin (K) and $k_{\text{B}} = 1.38 \times 10^{-23}\,\text{JK}^{-1}$ is known as Boltzmann's constant. An atom of a monatomic gas, for example, which has three degrees of freedom, as it can (only) drift in the x, y and z directions, has an average kinetic energy of $(3/2)k_{\text{B}}T$:

$T\,\text{K} = T\,°\text{C} + 273$

$$\langle E \rangle = \tfrac{1}{2}\,m\,\langle v^2 \rangle = \tfrac{3}{2}\,k_{\text{B}}T\,.$$

For neutrons, it has become conventional to equate the energy and temperature through the simplified relationship

$$E = k_{\text{B}}T, \qquad (1.20)$$

so that room temperature, $T = 293\text{K}$, corresponds to

$$\lambda = 1.80\,\text{Å} \quad \text{or} \quad v = 2.20\,\text{kms}^{-1} \quad \text{or} \quad E = 25.2\,\text{meV}\,,$$

or, for the spectroscopists, $204\,\text{cm}^{-1}$ in wavenumbers.

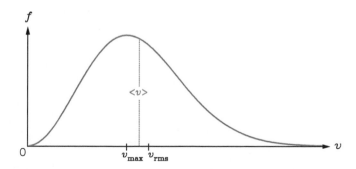

$$f(v) \propto v^2 \exp\left(-\frac{mv^2}{2\,k_{\text{B}}T}\right)$$

$$\Rightarrow v_{\text{max}}^2 = 2\,k_{\text{B}}T/m$$

$$\langle v \rangle^2 = \tfrac{8}{\pi}\,k_{\text{B}}T/m$$

$$\langle v^2 \rangle = 3\,k_{\text{B}}T/m$$

Fig. 1.12 The *Maxwell–Boltzmann* distribution of the frequency f of the speeds v of the atoms of a gas with mass m in thermal equilibrium at temperature T. It has a maximum at v_{max}, an average value of $\langle v \rangle$ and a root-mean-square of $v_{\text{rms}}^2 = \langle v^2 \rangle$.

1.6 A table of useful constants

Throughout this chapter we have made use of certain physical quantities, such as the mass of the neutron, the charge of the electron, Planck's constant and so on. We now list these, and other constants that will be encountered in this book, together in one place to provide a helpful reference.

Quantity	Symbol	Value (SI units)
Speed of light in a vacuum	c	2.997925×10^8 $\mathrm{m\,s^{-1}}$
Planck constant	h	6.626069×10^{-34} $\mathrm{J\,s}$
Planck constant $(h/2\pi)$	$\hbar$	1.054562×10^{-34} $\mathrm{J\,s}$
Boltzmann constant	k_B	1.380651×10^{-23} $\mathrm{J\,K^{-1}}$
Avogadro constant	N_A	6.022142×10^{23} $\mathrm{mol^{-1}}$
Gas constant	R	8.31447 $\mathrm{J\,K^{-1}\,mol^{-1}}$
Constant of gravitation	G	6.6742×10^{-11} $\mathrm{N\,m^2\,kg^{-2}}$
Permittivity of free space	ϵ_o	8.854189×10^{-12} $\mathrm{F\,m^{-1}}$
Permeability of free space	μ_o	1.256637×10^{-6} $\mathrm{H\,m^{-1}}$
Fine structure constant	α	7.297353×10^{-3}
Elementary charge	e	1.602177×10^{-19} C
Mass of an electron	m_e	9.109383×10^{-31} kg
Mass of a proton	m_p	1.672622×10^{-27} kg
Mass of a neutron	m_n	1.674927×10^{-27} kg
Classical radius of electron	r_e	2.817938×10^{-15} m
Bohr radius of atom	a_o	5.291771×10^{-11} m
Bohr magneton	μ_B	9.274009×10^{-24} $\mathrm{J\,T^{-1}}$
Nuclear magneton	μ_N	5.050783×10^{-27} $\mathrm{J\,T^{-1}}$

Waves, complex numbers and Fourier transforms

The theory of X-ray and neutron scattering relies heavily on the mathematics of waves. This chapter provides a tutorial introduction to the basic physical concepts, and the associated analytical tools, needed for an understanding of wave phenomena.

2.1 Sinusoidal waves

An everyday description of a wave would be a 'wiggle', or something that goes up-and-down as you move forward. The progression of the fluctuations could refer to changes in 'height' with respect to position at a fixed time, or with respect to time at a fixed position. Several examples of geometrical waves are shown in Fig. 2.1; they are unusual in that they have points where there are abrupt changes in the value of the function or its gradient. What they have in common with the more familiar *sinusoidal* variation of Fig. 2.2 is a regularly repeating pattern.

The *sine* and *cosine* curves of Fig. 2.2 are regarded as the archetypal waves, as they occur in many elementary physical situations; for example, the vibrations of an elastic string. Their smooth characteristics also make them amenable to analytical manipulation. An easy way of visualizing sinusoidal variations is to think about the projection of a circular motion onto the horizontal and vertical axes, as illustrated in Fig. 2.3.

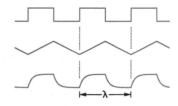

Fig. 2.1 Geometric examples of waves: square, triangular and exponential.

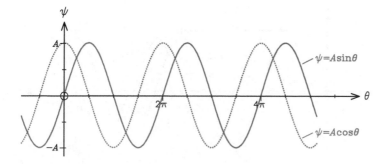

Fig. 2.2 The sinusoidal curves, or waves, $\psi = A \sin \theta$ and $\psi = A \cos \theta$.

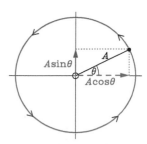

Fig. 2.3 The generation of sinusoidal variations through circular motion.

The two curves shown in Fig. 2.2 are identical apart from a lateral shift of $\pi/2$ radians, or $90°$: $\cos\theta = \sin(\theta + \pi/2)$. Hence, the general expression for a function of this type is

$$\psi = A\sin(\theta + \phi)\,, \tag{2.1}$$

where A is the *amplitude* of the wave and the angle ϕ, or the *phase*, controls its horizontal displacement with respect to $\sin\theta$. If the θ in eqn (2.1) varies linearly with position x, so that $\theta = kx$ where k is a constant, then we obtain a sinusoidal variation with respect to this physical coordinate:

$$\psi = A\sin(kx + \phi)\,. \tag{2.2}$$

Since the sine curve cycles around every 2π radians, the corresponding repeat distance, or *wavelength* λ, can be found from

$$\boxed{k = \frac{2\pi}{\lambda}}\,. \tag{2.3}$$

This is called the *wavenumber* and has SI units of $\mathrm{rad\,m^{-1}}$. Note that, as mentioned in Section 1.5, spectroscopists use the same term for $1/\lambda$ given in $\mathrm{cm^{-1}}$.

If the ϕ in eqn (2.2) itself varies linearly with time t, so that it can be written as $\phi = \phi_0 - \omega t$ where ϕ_0 and ω are constants, then we obtain the *travelling* wave

$$\psi = A\sin(kx - \omega t + \phi_0)\,. \tag{2.4}$$

That is to say, with $\omega > 0$, the sinusoidal variation in x moves steadily towards the right as time evolves; this is illustrated in Fig. 2.4. The crests and troughs of the translated wave will coincide with those of an earlier time after a duration T, called the *period*, given by

$$\boxed{\omega = \frac{2\pi}{T}}\,. \tag{2.5}$$

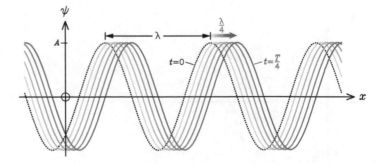

Fig. 2.4 The travelling wave of eqn (2.4) plotted as a function of x for several values of t, from zero to a quarter of the period.

The reciprocal of T, usually denoted by ν, is known as the *frequency* of the wave. It is related to its angular variant, ω, through

$$\omega = 2\pi\nu\,, \qquad (2.6)$$

$$\nu = \frac{1}{T}$$

with ω specified in $\mathrm{rad\,s}^{-1}$ and ν in cycles per second or Hz (hertz). The speed of the wave, c, follows readily from the observation that it moves forward by a distance λ in a time T:

$$c = \frac{\lambda}{T} = \frac{\omega}{k} = \nu\lambda\,, \qquad (2.7)$$

in agreement with the result quoted in eqn (1.16).

2.1.1 The direction of propagation

A negative prefactor was chosen for the ωt term in eqn (2.4) so that the wave would travel in the positive x direction; the opposite sign, $\psi = A\sin(kx + \omega t + \phi_{\rm o})$, gives a wave that moves backwards. In fact, a reversal also occurs with $\psi = A\sin(-kx - \omega t + \phi_{\rm o})$. Is there any reason for preferring one of these alternatives over the other to define the sense of the progression?

Conceptually, it would make more sense to associate the change of sign with the spatial term, rather than the temporal factor, because we are concerned with an orientation. This line of thought leads to the following generalization to accommodate fully the directional aspect of waves:

$$\mathbf{r} = (\,x\,,\,y\,,\,z\,)$$
$$\mathbf{k} = (k_x, k_y, k_z)$$

$$\psi = A\sin(\mathbf{k\cdot r} - \omega t + \phi_{\rm o})\,, \qquad (2.8)$$

$$\mathbf{k\cdot r} = k_x x + k_y y + k_z z$$

where the bold script $\mathbf{k}$ and $\mathbf{r}$ are *vectors*, and the dot between them indicates their 'scalar multiplication'. The vector $\mathbf{r}$ denotes a general position in space, with coordinates x, y and z, but what do the three components, k_x, k_y and k_z, of the *wavevector* $\mathbf{k}$ represent? Its magnitude, or *modulus*, $|\mathbf{k}| = k$ is the familiar wavenumber of eqn (2.3), and its orientation indicates the direction of propagation. For a wave travelling along the x direction, with $k_y = k_z = 0$, the scalar product $\mathbf{k\cdot r} = k_x x$ where $k_x = k$ for a forwards progression and $k_x = -k$ for the reverse.

$$|\mathbf{k}|^2 = k^2$$
$$\phantom{|\mathbf{k}|^2} = k_x^2 + k_y^2 + k_z^2$$

Since $\mathbf{r}$ and $\mathbf{k}$ are generally three-dimensional vectors, the wave of eqn (2.8) tends to be a function of x, y, z and t. As such, it represents a travelling 'plane wave' rather than a moving oscillation on a string. That is to say ψ, which could be the air pressure in a sound wave, is uniform in planes perpendicular to $\mathbf{k}$, but its value varies sinusoidally with time in the direction of the wavevector in accordance with the wavelength of eqn (2.3), the period of eqn (2.5) and the speed in eqn (2.7). The situation is illustrated for the two-dimensional analogue in Fig. 2.5.

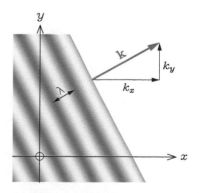

Fig. 2.5 The geometry of a plane wave.

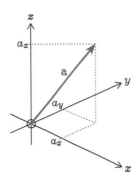

Magnitudes, directions and vectors

Quantities that have both a magnitude and a direction, such as a force, are called vectors. Unlike scalars, which only have a 'size', they cannot be quantified by a single number. They are defined by coordinates, or an array of numbers giving displacements with respect to a set of reference or *basis* axes. In the most common case of an x, y and z or *Cartesian* system, the vectors **a** and **b** can be written as

$$\mathbf{a} = (a_x, a_y, a_z) \quad \text{and} \quad \mathbf{b} = (b_x, b_y, b_z) .$$

Addition and subtraction are straightforward, in that the corresponding components are just combined separately:

$$\mathbf{a} + \mathbf{b} = (a_x + b_x, \, a_y + b_y, \, a_z + b_z) ,$$

with all the pluses replaced by minuses for a take away. The multiplication of a vector by a scalar, μ say, is also easy,

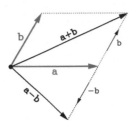

$$\mu \mathbf{a} = (\mu a_x, \mu a_y, \mu a_z) ,$$

and yields a vector with the original direction but an appropriately scaled length. The modulus, magnitude or length of a vector is given by *Pythagoras'* theorem; it's one for a *unit* or *normalized* vector.

A vector can be multiplied by another in two different ways. The first is a 'dot' or *scalar product*, which is a sum of the products of corresponding elements:

$$|\mathbf{a}|^2 = \mathbf{a} \cdot \mathbf{a} = a_x^2 + a_y^2 + a_z^2$$

$$\mathbf{a} \cdot \mathbf{b} = a_x b_x + a_y b_y + a_z b_z \tag{2.9}$$

$$\mathbf{a} \cdot \mathbf{b} = |\mathbf{a}| \, |\mathbf{b}| \cos \theta$$

and is geometrically the modulus of **a** times the modulus of **b** times the cosine of the angle between them. Vectors of non-zero length are perpendicular, or *orthogonal*, to each other if their dot product is zero; if they are also of unit length, they are said to be *orthonormal*.

Vectors can also be multiplied by a 'cross' or *vector product*. This is a bit more complicated since the result is a vector:

$$\mathbf{a} \times \mathbf{b} = (a_y b_z - b_y a_z, \, a_z b_x - b_z a_x, \, a_x b_y - b_x a_y) . \tag{2.10}$$

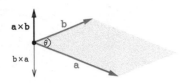

Geometrically, its magnitude is equal to the modulus of **a** times the modulus of **b** times the sine of the angle between them; this is also the area of the related parallelogram. The direction of the cross product is perpendicular to both **a** and **b**, and given by the 'right-hand screw rule': if the curl of the right-hand fingers indicates the sense of rotation needed to go from **a** to **b**, then the direction is given by the out-stretched thumb.

$$|\mathbf{a} \times \mathbf{b}| = |\mathbf{a}| \, |\mathbf{b}| \sin \theta$$

The physical meaning of a dot product holds irrespective of the dimensionality of the vectors and eqn (2.9) generalizes in an obvious way. The same is not true of a cross product, which is specific to a space of three dimensions (as considered here). The scalar product is also *symmetric* with respect to an interchange of **a** and **b** whereas the vector product is *antisymmetric*: the latter changes sign but the former does not. Division by a vector is not defined and must never be performed.

$$\mathbf{a} \cdot \mathbf{b} = \mathbf{b} \cdot \mathbf{a}$$

$$\mathbf{a} \times \mathbf{b} = -\mathbf{b} \times \mathbf{a}$$

2.1.2   The principle of superposition

A central feature of waves is that they pass through each other un-affected and, where overlapped, give a net result that is the sum of the individual contributions. This principle of *superposition* lies at the heart of scattering theory. Here we illustrate it with a couple of one-dimensional examples involving the combination of just two sinusoidal waves.

Consider first the case where the waves are identical but travel-ling in opposite directions. With the simplifying assignments that $A = 1$ and $\phi_\mathrm{o} = 0$ in eqn (2.4), to reduce the algebraic clutter, the principle of superposition yields

$$\psi = \sin(kx - \omega t) + \sin(-kx - \omega t)$$
$$= -2\sin(\omega t)\cos(kx),$$

$$\sin X + \sin Y = 2\sin\left(\tfrac{X+Y}{2}\right)\cos\left(\tfrac{X-Y}{2}\right)$$

where the second line follows from a trigonometric 'factor formula' and the antisymmetric properties of the sine function. This is called a *stationary* wave (Fig. 2.6), because there is no movement with time along the x direction. The separation of ψ into a product of spatial and temporal terms results in a purely 'up-and-down' oscillation, at a frequency of ω, with an amplitude that varies sinusoidally with wavelength $\lambda = 2\pi/k$. The locations at which the amplitude is zero are called *nodes*.

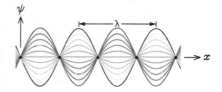

Fig. 2.6 A stationary wave, plotted as a function of x for several values of t.

As a second example, consider two waves travelling in the same direction with equal amplitudes but slightly different wavelengths and frequencies: $k \pm \Delta k$ and $\omega \pm \Delta\omega$, where the Δ-terms represent small departures from the average k and ω. With the simplification that $A = 1$ and $\phi_\mathrm{o} = 0$, as before, ψ is now a product of two travelling waves:

$$\psi = \sin\left([k+\Delta k]x - [\omega+\Delta\omega]t\right) + \sin\left([k-\Delta k]x - [\omega-\Delta\omega]t\right)$$
$$= -2\sin(kx - \omega t)\cos(\Delta k\,x - \Delta\omega\,t),$$

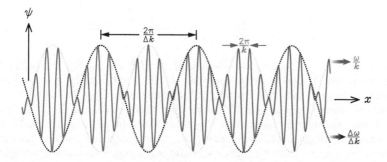

Fig. 2.7 The slowly varying 'beating' modulation, of wavelength $2\pi/\Delta k$, propagates with a speed of $\Delta\omega/\Delta k$, whereas the finer structure inside the envelope has the properties of the average wavelength and frequency, ω and k.

and is illustrated in Fig. 2.7. The amplitude of a sinusoid with the mean wavelength of $2\pi/k$, propagating with a speed ω/k, is modulated by a slowly varying envelope of wavelength $2\pi/\Delta k$, moving with a speed of $\Delta\omega/\Delta k$. This is the origin of the *beating* that is heard when neighbouring musical notes are played together: the sound becomes periodically louder and quieter.

Although we have only considered the combination of two similar waves, its generalization to the sum of many such components results in the formation of *wavepackets*; the beating modulation of Fig. 2.7 is just the most elementary example. The shape of the wavepacket will be preserved on propagation if all its constituents travel with the same speed c, when

$$\frac{\omega}{k} = c \quad \text{and} \quad \frac{d\omega}{dk} = c\,.$$

From Sivia and Rawlings (1999),
Foundations of Science Mathematics,
Oxford Chemistry Primers Series, **77**.

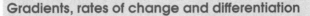

Gradients, rates of change and differentiation

The relationship between two quantities, x and y say, can be visualized with the aid of a graph. While the intersections of the associated curve with the x and y axes may be of interest, it is often more important to know the slope at any given point; that is, how quickly y increases, or decreases, as x changes, and vice versa. This issue is at the heart of the topic of *differentiation*, and the related rules and formulae are simply ways of calculating the gradient algebraically.

Let us begin with a precise definition of what is meant by the slope of a curve. Suppose that y is related to x through some function called 'f', usually written as $y = f(x)$, so that $f(x) = m\,x + c$ for a general straight line, and $f(x) = \sin(x)$ for a sinusoidal variation, and so on. Then, if the horizontal coordinate changes from x to $x + \Delta x$, where Δx represents a small increment, the value of y is altered from $f(x)$ to $f(x+\Delta x)$. The gradient, at a point x, is defined to be the ratio of the change in the vertical coordinate, Δy, to that of the horizontal increment, as Δx becomes vanishingly small. This can be stated formally as

$$\frac{dy}{dx} = \lim_{\Delta x \to 0} \frac{\Delta y}{\Delta x} = \lim_{\Delta x \to 0} \frac{f(x + \Delta x) - f(x)}{\Delta x}\,, \tag{2.11}$$

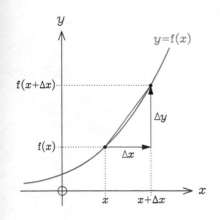

$$y' = \frac{dy}{dx} = \frac{d}{dx}(y) = f'(x)$$

$$y'' = \frac{d^2y}{dx^2} = \frac{d}{dx}\left(\frac{dy}{dx}\right) = f''(x)$$

where dy/dx is known as the *derivative*, or *differential coefficient*, and is pronounced 'dy-by-dx'. The tendency of $\Delta x \to 0$ has to be approached gradually to ascertain the *limiting* value of the ratio $\Delta y/\Delta x$, as both increments are individually equal to zero when the condition is met. Strictly speaking, we should check that the same value of dy/dx is obtained whether Δx is positive or negative, but this is assured as long as the curve $y = f(x)$ is 'smooth'; inconsistencies will arise if kinks and sudden breaks (or discontinuities) are present, and the function is said to be *nondifferentiable* at those points.

If the speed of the sinusoidal waves varies with their wavelength, because the frequency does not happen to be directly proportional to the wavenumber in the medium of interest, then the wavepacket will change with time. This phenomenon is called *dispersion*, and the relationship between ω and k which determines the nature of the 'spreading',

$$\omega = \omega\left(k\right),$$

is called the 'dispersion relation' or the 'dispersion curve' (Fig. 2.8). For the non-dispersive case, $\omega = c\,k$, there is a unique speed, c, associated with the propagation of a wavepacket. The ratio ω/k and the derivative $\mathrm{d}\omega/\mathrm{d}k$ still yield useful characteristic speeds, however, when there is a dominant contribution from sinusoidal waves around a particular wavelength:

$$v_\phi = \frac{\omega}{k} \quad \text{and} \quad v_\mathrm{g} = \frac{\mathrm{d}\omega}{\mathrm{d}k}, \tag{2.12}$$

where v_ϕ is called the *phase velocity*, and gives the rate at which the crests and troughs of the local wavefront move, and v_g is the *group velocity*, which indicates how fast the envelope of the wavepacket travels.

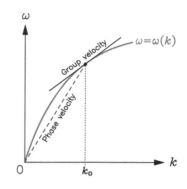

Fig. 2.8 A dispersion curve, and the related phase and group velocities for waves in the neighbourhood of k_o.

2.2 Complex numbers

The analysis of wave phenomena is aided greatly by the use of *complex numbers*. In particular, by a result which links an exponential to sines and cosines:

$$\boxed{\mathrm{e}^{\mathrm{i}\theta} = \cos\theta + \mathrm{i}\sin\theta}, \tag{2.13}$$

where $\mathrm{i}^2 = -1$. Since complex numbers play a central role in theoretical work, we will devote a few pages to them; as with most of the mathematical background given in this book, the material is based on Sivia and Rawlings (1999).

2.2.1 Definition

If any number, integer or fraction, positive or negative, is multiplied by itself, then the result is always greater than, or equal, to zero. What, then, is the square root of -9? To address this question we need to invent an *imaginary* number, usually denoted by 'i', whose square is defined to be negative:

$$\mathrm{i}^2 = -1. \tag{2.14}$$

A *real* number, say b (where $b^2 \geqslant 0$), times i is also imaginary; it's just b times bigger than i. If a is also an ordinary number, then the sum z of a and ib,

$$z = a + \mathrm{i}b, \tag{2.15}$$

$$\sqrt{-9} = \pm 3\mathrm{i}$$

is known as a 'complex' number; this does not indicate an intrinsic difficulty with the concept, but highlights the hybrid nature of the entity. It consists of both a real part and an imaginary one:

$$\mathcal{R}e\{z\} = a \quad \text{and} \quad \mathcal{I}m\{z\} = b. \tag{2.16}$$

It may seem odd that $\mathcal{I}m\{z\}$ is b rather than $\mathrm{i}b$, but this is because it represents the size of the imaginary component.

2.2.2 Basic algebra

To add or subtract complex numbers, we simply add or subtract the real and imaginary parts separately:

$$a + \mathrm{i}b \pm (c + \mathrm{i}d) = a \pm c + \mathrm{i}(b \pm d), \tag{2.17}$$

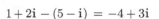

where a, b, c and d are real. The usual rules of algebra apply for brackets and multiplication, except that every occurrence of i^2 is replaced by -1. Thus, it's easy to show that the product of $a + \mathrm{i}b$ and $c + \mathrm{i}d$ is given by

$$(a + \mathrm{i}b)(c + \mathrm{i}d) = ac - bd + \mathrm{i}(ad + bc), \tag{2.18}$$

$(1 + 2\mathrm{i})(3 - \mathrm{i}) = 5 + 5\mathrm{i}$

since $\mathrm{i}^2 bd = -bd$. Division involves the use of a *complex conjugate*, so let us consider this first.

The conjugate of a complex number z, denoted by z^*, is defined to have the same real part but the opposite imaginary component; that is, $\mathcal{R}e\{z^*\} = \mathcal{R}e\{z\}$ and $\mathcal{I}m\{z^*\} = -\mathcal{I}m\{z\}$. In terms of eqn (2.15), therefore,

$$z^* = (a + \mathrm{i}b)^* = a - \mathrm{i}b. \tag{2.19}$$

Hence, complex numbers and their conjugates satisfy the following relationships:

$$\begin{aligned} z + z^* &= & 2a & = 2\mathcal{R}e\{z\} \\ z - z^* &= & 2\mathrm{i}b & = 2\mathrm{i}\,\mathcal{I}m\{z\} \\ z\,z^* &= & a^2 + b^2 & = |z|^2 \end{aligned} \tag{2.20}$$

We will come to the meaning of $|z|$ shortly, but the important point about eqn (2.20) is that the product $z\,z^*$ is a real number. This feature enables us to calculate the real and imaginary part of the ratio of two complex numbers by multiplying both the top and bottom by the conjugate of the denominator

$$\frac{a + \mathrm{i}b}{c + \mathrm{i}d} = \frac{a + \mathrm{i}b}{c + \mathrm{i}d} \times \frac{c - \mathrm{i}d}{c - \mathrm{i}d} = \frac{ac + bd + \mathrm{i}(bc - ad)}{c^2 + d^2}. \tag{2.21}$$

To evaluate the ratio $(1+2\mathrm{i})/(3-\mathrm{i})$, for example, we multiply it by unity in the form $(3+\mathrm{i})/(3+\mathrm{i})$; this gives a real denominator of 10, and a complex numerator of $1+7\mathrm{i}$. Hence the result is $1/10 + \mathrm{i}\,7/10$.

2.2.3 The Argand diagram

So far we have considered complex numbers from an algebraic point of view; it is often helpful to think of them in geometrical terms. This is easily done with the aid of an *Argand diagram* where the horizontal, or x, axis of a graph is seen as representing the real part of a complex number, and the vertical, or y, axis gives the imaginary component. Thus the point with (x, y) coordinates (a, b) corresponds to the complex number $a + ib$. Its conjugate z^* is a reflection in the real axis. An alternative way of specifying the location of a point on a graph is through its distance r from the origin, and the anticlockwise angle θ that this 'radius' makes with the (positive) real axis. In this system r is known as the *modulus*, magnitude or amplitude of z; θ is called its *argument* or phase.

The quantities r, θ, a and b in the Argand diagram are related through elementary trigonometry by

$$a = r \cos \theta \quad \text{and} \quad b = r \sin \theta, \quad (2.22)$$

or, in the reverse sense, by

$$r^2 = a^2 + b^2 \quad \text{and} \quad \theta = \tan^{-1}(b/a). \quad (2.23)$$

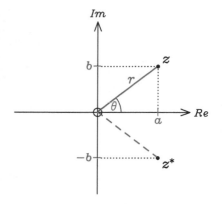

A comparison between eqns (2.20) and (2.23) shows that $z\,z^* = r^2$, where $r = |z|$ is the modulus of the complex number. The second part of eqn (2.23) needs qualification because there is an ambiguity of $180°$ with $\tan^{-1}(b/a)$. The arctangent of unity, for example, could be either $45°$ or $-135°$. For complete consistency with eqn (2.22), $0 < \theta < \pi$ (radians) for $b > 0$ and $-\pi < \theta < 0$ for $b < 0$; θ is zero for $a > 0$, and $\pm\pi$ for $a < 0$, if $b = 0$. It is also worth remembering that θ is only defined to within a factor of 2π, because we could add (or subtract) any integer number of $360°$ to it and obtain the same point in the Argand diagram.

2.2.4 The imaginary exponential

Perhaps the most important result in complex analysis concerns the exponential of an imaginary number:

$$e^{i\theta} = \cos\theta + i\sin\theta, \quad (2.24)$$

$$e^x = 1 + x + \frac{x^2}{2!} + \frac{x^3}{3!} + \cdots$$

where θ is in radians. This equation can be verified by substituting $x = i\theta$ in the *Taylor series* expansion for e^x, and collecting the odd and even powers of θ separately; remembering that $i^2 = -1$, a comparison with the Taylor series for $\sin\theta$ and $\cos\theta$ yields eqn (2.24). The product of r and $e^{i\theta}$ allows a complex number to be expressed in a very compact form in terms of its modulus and argument:

$$\sin\theta = \theta - \frac{\theta^3}{3!} + \frac{\theta^5}{5!} - \cdots$$

$$\cos\theta = 1 - \frac{\theta^2}{2!} + \frac{\theta^4}{4!} - \cdots$$

$$z = a + ib = r\left(\cos\theta + i\sin\theta\right) = r\,e^{i\theta}, \quad (2.25)$$

where a, b, r and θ are related through eqns (2.22) and (2.23). As can be seen from the Argand diagram, and verified by the symmetry properties of sines and cosines, its conjugate entails the replacement of θ with $-\theta$:

$$z^* = a - \mathrm{i}b = r\left(\cos\theta - \mathrm{i}\sin\theta\right) = r\,\mathrm{e}^{-\mathrm{i}\theta}, \qquad (2.26)$$

from which the result $z\,z^* = r^2$ follows immediately.

Although the exponential form of a complex number is very useful when dealing with roots and logarithms, and provides a valuable insight into products and quotients, our interest here is in its relationship with waves. This hinges on eqn (2.24), which enables eqn (2.8) to be written as the imaginary part of

$$\psi = A\,\mathrm{e}^{\mathrm{i}(\mathbf{k}\bullet\mathbf{r}-\omega t)}, \qquad (2.27)$$

where A is now a complex number whose modulus and argument give the amplitude and phase offset of the wave, respectively:

$$A = |A|\,\mathrm{e}^{\mathrm{i}\phi_\mathrm{o}}. \qquad (2.28)$$

The real part of eqn (2.27) also represents the same wave, apart from a difference of $90°$ in the value of ϕ_o.

From Sivia and Rawlings (1999), *Foundations of Science Mathematics*, Oxford Chemistry Primers Series, **77**.

$$a_n = \frac{1}{n!}\left.\frac{\mathrm{d}^n f}{\mathrm{d}x^n}\right|_{x_\mathrm{o}}$$

Taylor series

When dealing with a complicated function, it can be useful to approximate it with one of a simpler form. While the latter may not represent a complete and accurate description of the situation at hand, it frequently provides the only means of making analytical progress. There are many approximations that could be used, of course, but it is the one that captures the salient features that is most helpful. A Taylor series is appropriate when our principal interest lies in the behaviour of a function in the neighbourhood of a particular point.

Consider the curve $y=\mathrm{f}(x)$. The crudest approximation to this function is a horizontal line $y=a_0$, where a_0 is a constant; if $a_0=\mathrm{f}(x_\mathrm{o})$, then it will even be correct at $x=x_\mathrm{o}$. A better approximation would be a sloping line $y = a_0 + a_1(x-x_\mathrm{o})$, where the coefficient a_1 allows for a non-zero gradient. Continuing along this path, we could add a quadratic (or curvature) term $a_2(x-x_\mathrm{o})^2$, a cubic contribution $a_3(x-x_\mathrm{o})^3$, and so on, to gain further improvements. Thus, a function $\mathrm{f}(x)$ can be approximated about the point x_o by using a *polynomial* expansion:

$$\mathrm{f}(x) \approx a_0 + a_1(x-x_\mathrm{o}) + a_2(x-x_\mathrm{o})^2 + a_3(x-x_\mathrm{o})^3 + \cdots. \qquad (2.29)$$

This is the essence of a Taylor series. Its advantage is that the right-hand side of eqn (2.29) is usually easier to calculate, differentiate, integrate, and generally manipulate, than the expression on the left. The case of $x_\mathrm{o}=0$, when the Taylor series simplifies, is called a *Maclaurin series*.

The benefit of using eqn (2.27) over (2.8) in wave analysis is that exponentials are easier to deal with mathematically than sinusoids; multiplication, differentiation and *integration*, for example, are more straightforward. As an illustration of this advantage, let's derive the 'compound angle' formulae for sines and cosines with complex numbers. Starting with the rule of eqn (1.2) for combining powers,

$$e^{i(\alpha+\beta)} = e^{i\alpha}\, e^{i\beta}\,,$$

and expanding the exponentials with eqn (2.24),

$$\cos(\alpha+\beta) + i\sin(\alpha+\beta) = (\cos\alpha + i\sin\alpha)\,(\cos\beta + i\sin\beta)\,,$$

the equating of the real and imaginary parts on the left- and right-hand sides yields the desired results:

$$\cos(\alpha+\beta) = \cos\alpha\,\cos\beta - \sin\alpha\,\sin\beta\,, \qquad (2.30)$$

$$\sin(\alpha+\beta) = \sin\alpha\,\cos\beta + \cos\alpha\,\sin\beta\,. \qquad (2.31)$$

As well as being the real and imaginary parts of $\exp(i\theta)$, sines and cosines can also be expressed as

$$\cos\theta = \frac{e^{i\theta} + e^{-i\theta}}{2} \quad \text{and} \quad \sin\theta = \frac{e^{i\theta} - e^{-i\theta}}{2i}\,, \qquad (2.32)$$

which follow from the addition and subtraction, respectively, of eqn (2.24) with its complex conjugate.

2.3   Fourier series

Let's begin wave analysis by considering how periodic signals, such as those in Fig. 2.1, can be decomposed into the sum of sinusoids. Suppose that the function $f(x)$ repeats itself after a 'distance' of λ, so that

$$f(x) = f(x+\lambda)\,. \qquad (2.33)$$

This has the same periodicity as sines and cosines of wavenumber $k = 2\pi/\lambda$. A simple approximation to $f(x)$, which matches its wavelength, is therefore

$$f(x) \approx a_0 + a_1\cos(kx) + b_1\sin(kx)\,, \qquad (2.34)$$

where a_0, a_1 and b_1 are constants whose values need to be selected in some way. The crudest assignment would be to set both a_1 and b_1 equal to zero, giving an invariant $f(x) \approx a_0$, but the *linear* combination of $\sin(kx)$ and $\cos(kx)$ allows for a sinusoidal variation with the correct period and an appropriate amplitude and phase:

$$a\cos(kx) + b\sin(kx) = A\sin(kx+\phi)\,,$$

where $a = A\sin\phi$ and $b = A\cos\phi$, in accordance with eqn (2.31).

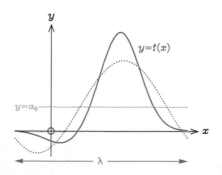

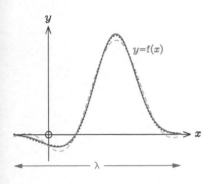

$$\sin(\theta) = -\sin(-\theta)$$
$$\cos(\theta) = \cos(-\theta)$$

The sines and cosines of $2\,kx$, $3\,kx$, $4\,kx$, and so on, also satisfy the periodicity of eqn (2.33); they just go through several, or many, complete cycles in the interval λ. We can obtain a better approximation to $f(x)$, therefore, by including contributions from these higher-order terms:

$$f(x) \approx a_0 + a_1 \cos(kx) + a_2 \cos(2\,kx) + a_3 \cos(3\,kx) + \cdots$$
$$+ b_1 \sin(kx) + b_2 \sin(2\,kx) + b_3 \sin(3\,kx) + \cdots \tag{2.35}$$

This expansion is called a *Fourier series*, and eqn (2.34) is simply the first-order version of it which contains only the lowest, or fundamental, harmonic.

We will come to the evaluation of the coefficients a_n and b_n, for integer n, shortly but note that one of the sets goes to zero if $f(x)$ possesses a symmetry about the y-axis:

$$f(x) = \begin{cases} f(-x) & \implies b_n = 0 \\ -f(-x) & \implies a_n = 0 \end{cases} \tag{2.36}$$

because sines and cosines are *odd* and *even* functions, respectively. The generalization of eqn (2.35) explains why the invariant term is designated as a_0, and why there is no corresponding b_0 (apart from its general redundancy): they are the coefficients of $\cos(0) = 1$ and $\sin(0) = 0$, with the b_0 being unnecessary since it adds nothing to the Fourier series.

2.3.1 Orthogonality and the Fourier coefficients

A prescription for the a_n and b_n in eqn (2.35) presents itself once we realize that the related sine and cosine functions are *orthogonal*. By this we mean that the *integral* of the product of any two over the interval of the period λ will be zero, unless they happen to be exactly the same functions:

$$\int_0^\lambda \sin(mkx) \sin(nkx)\, dx = \begin{cases} 0 & \text{if } m \neq n, \\ \frac{\lambda}{2} & \text{if } m = n, \end{cases} \tag{2.37}$$

with an identical expression for $\cos(mkx)\cos(nkx)$, but $n \neq 0$, and

$$\int_0^\lambda \sin(mkx) \cos(nkx)\, dx = 0. \tag{2.38}$$

Although these sines and cosines aren't perpendicular in a geometrical sense, this type of integral is the functional analogue of a dot product which is zero for orthogonal vectors.

If we multiply eqn (2.35) through by one of the sine or cosine functions, $\sin(mkx)$ or $\cos(mkx)$, and integrate the resultant products over the period λ, then all but one of the terms on the right-hand side will be zero due to eqns (2.37) and (2.38). The surviving $m = n$ contributions yield the formulae for the Fourier coefficients:

$$a_n = \tfrac{2}{\lambda} \int_0^\lambda f(x) \cos(n\,kx)\,\mathrm{d}x \quad \text{and} \quad b_n = \tfrac{2}{\lambda} \int_0^\lambda f(x) \sin(n\,kx)\,\mathrm{d}x \qquad (2.39)$$

for $n = 1, 2, 3, \ldots$, from which eqn (2.36) can be verified. If eqn (2.35) is integrated over the period λ as it stands, then the constant a_0 is seen to be the average value of $f(x)$:

$$\int_0^\lambda \sin(n\,kx)\,\mathrm{d}x = 0$$

$$a_0 = \tfrac{1}{\lambda} \int_0^\lambda f(x)\,\mathrm{d}x\;. \qquad (2.40)$$

$$\int_0^\lambda \mathrm{d}x = \lambda$$

From Sivia and Rawlings (1999), *Foundations of Science Mathematics*, Oxford Chemistry Primers Series, **77**.

Cumulative properties and integrals

While differentiation is concerned with the slope of $y = f(x)$, integration deals with the 'area under the curve'. This relates to the average and cumulative behaviour of y, over some range in x.

To set up a definition of an integral, consider the region bounded by the straight lines $x = a$, $x = b$ and $y = 0$, and the curve $y = f(x)$. The size of the enclosure can be estimated by approximating it as a whole series of narrow vertical strips, and adding together the areas of these contiguous rectangular blocks. If the x-axis between a and b is divided into N equal intervals, then the width of each strip is given by $\Delta x = (b-a)/N$; the corresponding heights of the thin blocks are equal to the values of the function $f(x)$ at their central positions. In other words, the area of the jth strip, which is at $x = x_j$ and of height $y = f(x_j)$, is $f(x_j)\,\Delta x$; the index j ranges from 1 to N, of course, with $x_1 = a + \Delta x/2$ and $x_N = b - \Delta x/2$. As N tends to infinity, $\Delta x \to 0$ and the approximation to the area under the curve becomes ever more accurate. This limiting form of the summation procedure defines an integral

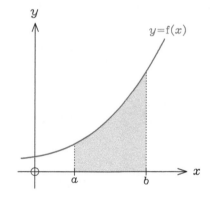

$$\int_a^b y\,\mathrm{d}x = \int_a^b f(x)\,\mathrm{d}x = \lim_{N\to\infty} \sum_{j=1}^{N} f(x_j)\,\Delta x\,, \qquad (2.41)$$

where the symbol $\int \mathrm{d}x$ is read as the 'integral, from a to b, with respect to x'. The use of the term 'area' in the above discussion needs some qualification, in that it can be negative; this is because the 'height' of a strip $f(x_j) < 0$ whenever the curve $y = f(x)$ lies below the x-axis (and even the 'width' $\Delta x < 0$ if $b < a$).

Although an integral is defined as the limiting form of a summation, it is usually calculated analytically by noting that 'integration is the reverse of differentiation'. While this may not be obvious, it is easily illustrated with an example from everyday kinematics: the distance travelled by a car (say) is the integral of the speed with respect to time, and speed is the rate of change of distance with time (a derivative).

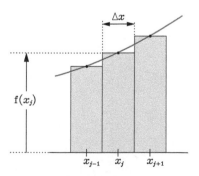

2.3.2 The complex Fourier series

The Fourier series of eqn (2.35) can be written in a very compact form by using complex numbers:

$$\mathrm{f}(x) \;=\; \sum_{n=-\infty}^{\infty} c_n \, \mathrm{e}^{inkx}, \qquad (2.42)$$

where the Σ stands for a summation over integer values of n, from $-\infty$ to ∞, and the approximation has been replaced by an equality to denote a definition. The right-hand side of eqn (2.42) will yield a real function, $\mathrm{f}(x)$, as long as the complex coefficients c_n satisfy the conjugacy condition

$$c_{-n} \;=\; c_n^{*}. \qquad (2.43)$$

This follows from eqn (2.20) because the contribution to the sum from pairs of positive and negative values of n of equal magnitudes will then be

$$\left(c_n \, \mathrm{e}^{inkx} \right)^{*} = c_n^{*} \, \mathrm{e}^{-inkx}$$

$$c_n \, \mathrm{e}^{inkx} + c_{-n} \, \mathrm{e}^{-inkx} \;=\; 2\,\mathcal{Re}\left\{ c_n \, \mathrm{e}^{inkx} \right\}$$

$$=\; a_n \cos(nkx) + b_n \sin(nkx)\,,$$

for $n \neq 0$, where we have substituted $2\,c_n = a_n - ib_n$ in the second line to obtain consistency with eqn (2.35). In fact, the formula for the complex coefficients is simply

$$c_n \;=\; \tfrac{1}{\lambda} \int_{0}^{\lambda} \mathrm{f}(x)\, \mathrm{e}^{-inkx} \, \mathrm{d}x\,, \qquad (2.44)$$

with $c_0 = a_0$.

2.4 Fourier transforms

We began our discussion of Fourier series by considering how a periodic function could be decomposed into, or approximated by, a sum of sinusoidal waves. The analysis can be extended to the non-periodic case by letting $\lambda \to \infty$, so that no repetitions are required of $\mathrm{f}(x)$ within a finite interval. To carry out this limiting procedure, it is helpful to define

$$\Delta k \;=\; \frac{2\pi}{\lambda} \qquad \text{and} \qquad k_n = n\,\Delta k$$

because the wavenumber of the fundamental harmonic, Δk, shrinks gradually to zero as λ gets ever larger and k_n approaches a continuum even with integer n. The imaginary exponentials of eqns (2.42) and (2.44) can then be written as

$$\mathrm{e}^{ik_n x} \qquad \text{and} \qquad \mathrm{e}^{-ik_n x}\,,$$

and the coefficients expressed as

$$c_n = \alpha \, \mathrm{F}(k_n) \, \Delta k \,,$$

where α is a constant and $\mathrm{F}(k)$ is a continuous function of k. With these substitutions, eqns (2.42) and (2.44) become

$$\mathrm{f}(x) = \alpha \sum_{n=-\infty}^{\infty} \mathrm{F}(k_n) \, \mathrm{e}^{\mathrm{i} k_n x} \Delta k \quad \text{and} \quad \mathrm{F}(k_n) = \frac{1}{2\pi\alpha} \int_{-\lambda/2}^{\lambda/2} \mathrm{f}(x) \, \mathrm{e}^{-\mathrm{i} k_n x} \, \mathrm{d}x \,,$$

where we have expressed the integral over a period as being from $-\lambda/2$ to $\lambda/2$, instead of 0 to λ, for a more symmetrical appearance. The limit of $\lambda \to \infty$, when $\Delta k \to 0$, can now be taken safely, and yields the integrals

$$\mathrm{f}(x) = \frac{1}{\sqrt{2\pi}} \int_{-\infty}^{\infty} \mathrm{F}(k) \, \mathrm{e}^{\mathrm{i} k x} \, \mathrm{d}k \qquad (2.45)$$

and

$$\mathrm{F}(k) = \frac{1}{\sqrt{2\pi}} \int_{-\infty}^{\infty} \mathrm{f}(x) \, \mathrm{e}^{-\mathrm{i} k x} \, \mathrm{d}x \qquad (2.46)$$

as being the continuum versions of eqns (2.42) and (2.44), where we have set $\alpha = 1/\sqrt{2\pi}$ for aesthetic reasons of symmetry. These come as a linked pair, and define a *Fourier transform* and its *inverse*; which one is called which is quite arbitrary.

While the exponents of a Fourier transform and its inverse must have opposite signs, their precise definitions are a matter of convention; the choice of α is up to us, for example. If the wavenumber is taken to be $1/\lambda$ instead of $2\pi/\lambda$, as done by spectroscopists, then the exponents will be $\pm \mathrm{i} 2\pi k x$ and neither integral will have a scaling term:

$$\mathrm{f}(x) = \int_{-\infty}^{\infty} \mathrm{F}(k) \, \mathrm{e}^{\mathrm{i} 2\pi k x} \, \mathrm{d}k \qquad \text{and} \qquad \mathrm{F}(k) = \int_{-\infty}^{\infty} \mathrm{f}(x) \, \mathrm{e}^{-\mathrm{i} 2\pi k x} \, \mathrm{d}x \,,$$

where k is measured in cycles per unit length, typically cm^{-1}, rather than the SI radians per metre.

Although our goal is to gain a physical insight into Fourier transforms, we first need to discuss some of their formal properties. Basic symmetries are a good place to start, as the most common one is the continuum analogue of eqn (2.43):

$$\mathrm{f}(x) = \mathrm{f}(x)^{*} \quad \Longleftrightarrow \quad \mathrm{F}(-k) = \mathrm{F}(k)^{*} \,, \qquad (2.47)$$

which states that the Fourier transform of a real function is 'conjugate symmetric'. If one of them possesses a symmetry about the origin, then so too will the other:

$$f(x) = \begin{cases} f(-x) \\ -f(-x) \end{cases} \iff F(k) = \begin{cases} F(-k)\,, \\ -F(-k)\,. \end{cases} \tag{2.48}$$

Equations (2.47) and (2.48) can be combined to show that the Fourier transform of a real and symmetric function is also real and even, whereas that of a real and antisymmetric function is imaginary and odd; this is equivalent to eqn (2.36).

The substitution of $k=0$ in eqns (2.46) and (2.47) reveals F(0) to be proportional to the area under the curve $y=f(x)$,

$$F(0) = \tfrac{1}{\sqrt{2\pi}} \int\limits_{-\infty}^{\infty} f(x)\,\mathrm{d}x\,, \tag{2.49}$$

and necessarily real if $f(x)=f(x)^{*}$. It will equal zero if $f(x)=-f(-x)$. Technically, the integral of the modulus, $|f(x)|$, must be bounded (or finite) if its Fourier transform is to exist everywhere; this is known as the *Dirichlet* condition.

2.4.1 Convolution theorem

One of the most useful results in Fourier theory concerns the *convolution* of two functions. Mathematically, the convolution of $g(x)$ and $h(x)$ is defined by

$$g(x) \otimes h(x) = \int\limits_{-\infty}^{\infty} g(t)\,h(x-t)\,\mathrm{d}t\,, \tag{2.50}$$

where $g\otimes h$ is read as 'g convolved with h', and physically represents a 'blurring' of $g(x)$ by $h(x)$. This can be understood from the example of Fig. 2.9, where $g(x)$ consists of four spikes, or δ-*functions*, and $h(x)$ is a broad asymmetric function. The convolution is carried out

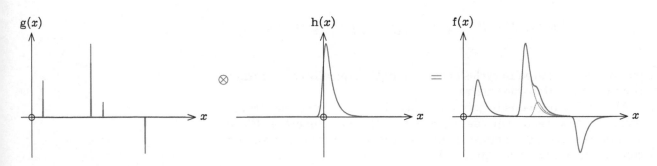

Fig. 2.9 The convolution of the spiky function $g(x)$ with the broad asymmetric function $h(x)$: $f(x) = g(x) \otimes h(x)$.

by replacing each of the the sharp peaks in $g(x)$ with scaled copies of $h(x)$ and adding together the four contributions; those from the two closely spaced components in the middle, shown by dotted grey lines, combine to give a resultant function where the constituent doublet is no longer resolved clearly. Although it's not as easy to visualize it the other way around, eqn (2.50) can equally be thought of as the blurring of $h(x)$ by $g(x)$.

$$g(x) \otimes h(x) = h(x) \otimes g(x)$$

The convolution theorem states that the Fourier transform of the convolution of two functions is proportional to the product of their Fourier transforms:

$$\boxed{f(x) = g(x) \otimes h(x) \quad \Longleftrightarrow \quad F(k) = \sqrt{2\pi}\, G(k) \times H(k)} \;, \qquad (2.51)$$

where $F(k)$, $G(k)$ and $H(k)$ are the Fourier transforms of $f(x)$, $g(x)$ and $h(x)$, respectively, according to eqn (2.46). Given the reciprocity between a Fourier transform and its inverse,

$$f(x) = g(x) \times h(x) \quad \Longleftrightarrow \quad F(k) = \tfrac{1}{\sqrt{2\pi}}\, G(k) \otimes H(k)\,. \qquad (2.52)$$

The power of eqn (2.51) will be illustrated in a physical sense in the next section, and throughout this book, but its computational benefit stems from the fact that it's much easier to multiply functions than to convolve them. To work out $g(x) \otimes h(x)$ numerically, for example, it's quicker to use a *fast Fourier transform* (FFT) computer subroutine to calculate $G(k)$ and $H(k)$, and inverse Fourier transform their product, than to compute the integral of eqn (2.50) directly.

$$g(x) \otimes h(x) = $$
$$\tfrac{1}{2\pi} \int\limits_{-\infty}^{\infty} G(k)\,H(k)\,\mathrm{e}^{ikx}\,\mathrm{d}k$$

Putting $k=0$ in eqn (2.51), and interpreting $F(0)$, $G(0)$ and $H(0)$ with eqn (2.49), shows that the area under the convolution is equal to the product of the corresponding individual integrals:

$$\int\limits_{-\infty}^{\infty} \big[\,g(x) \otimes h(x)\,\big]\,\mathrm{d}x \;=\; \int\limits_{-\infty}^{\infty} g(x)\,\mathrm{d}x \;\times\; \int\limits_{-\infty}^{\infty} h(t)\,\mathrm{d}t\,. \qquad (2.53)$$

This can be seen from the example of Fig. 2.10, where an array of different shaped peaks is convolved with a Gaussian. Although the

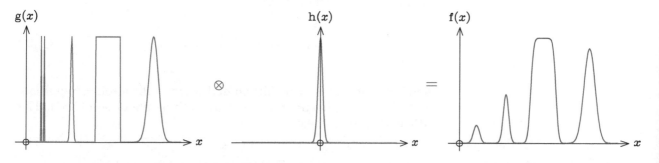

Fig. 2.10 The convolution of a function with an array of different shaped peaks, $g(x)$, with a Gaussian, $h(x)$.

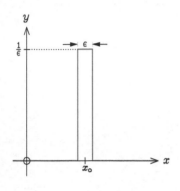

$$\delta(x-x_\mathrm{o}) \otimes \mathrm{h}(x) = \mathrm{h}(x-x_\mathrm{o})$$

The Dirac δ-function

A Dirac δ-function, $\delta(x-x_\mathrm{o})$, is a sharp spike of unit area at a given location, x_o; its simplicity as a 'point impulse' makes it a useful test object for studying equations that model physical situations. Mathematically, it is defined by

$$\delta(x-x_\mathrm{o}) = 0 \ \text{ if } x \neq x_\mathrm{o} \quad \text{and} \quad \int_{-\infty}^{\infty} \delta(x-x_\mathrm{o}) \, \mathrm{d}x = 1 \,,$$

and can be thought of as the limiting form of a variety of functions as they become ever narrower. Of these the most straightforward is a rectangular column of width ϵ, centred on $x = x_\mathrm{o}$, and height $1/\epsilon$; this acquires the properties of $\delta(x-x_\mathrm{o})$ in the limit of $\epsilon \to 0$. An important corollary of the above definition is

$$\int_{a}^{b} \mathrm{f}(x) \, \delta(x-x_\mathrm{o}) \, \mathrm{d}x = \begin{cases} \mathrm{f}(x_\mathrm{o}) & \text{if } a < x_\mathrm{o} < b \,, \\ 0 & \text{otherwise} \,, \end{cases} \tag{2.54}$$

so that integrals involving a δ-function are easy to evaluate.

two spikes on the left of $\mathrm{g}(x)$ merge into one in $\mathrm{f}(x)$, because they are very closely spaced compared with the width of $\mathrm{h}(x)$, the areas of the various components in the blurred output are proportional to those of the input signal. The amplitudes of the narrowest peaks are affected the most, since their relative spreading is the greatest as a result of the convolution; the slowly varying parts of the structure change the least.

2.4.2 Auto-correlation function

The last Fourier concept that we need to consider concerns the *auto-correlation function*, or ACF, which provides information on the distance distribution of the various structures in $\mathrm{f}(x)$. Mathematically, the ACF of $\mathrm{f}(x)$ is defined by

$$\mathrm{ACF}(x) = \int_{-\infty}^{\infty} \mathrm{f}(t)^{*} \, \mathrm{f}(x+t) \, \mathrm{d}t \,, \tag{2.55}$$

and is real if $\mathrm{f}(x)^{*} = \mathrm{f}(x)$. Although this looks like a self-convolution, or $\mathrm{f}(x)^{*} \otimes \mathrm{f}(-x)$, it's not the best way to think about eqn (2.55). The ACF is largest at the origin,

$$\mathrm{f}(x)^{*} \mathrm{f}(x) = \left| \mathrm{f}(x) \right|^{2} \geqslant 0$$

$$\mathrm{ACF}(0) = \int_{-\infty}^{\infty} \mathrm{f}(t)^{*} \, \mathrm{f}(t) \, \mathrm{d}t \,,$$

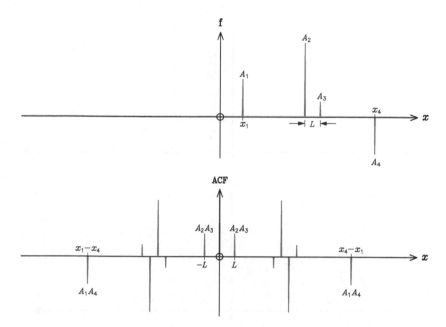

Fig. 2.11 An $f(x)$ consisting of four sharp peaks and its auto-correlation function. The spike at the origin of the ACF should be three times higher than drawn, and has been suppressed for clarity. The relationship of the closest and farthest peaks in $f(x)$ to their corresponding mutual contributions in the ACF is indicated.

because everything correlates with itself. The value of the ACF at a distance L away from the origin is calculated by multiplying $f(x)$ with a copy that's displaced by L relative to it, $f(L+x)$, and integrating the product; its magnitude is a measure of how much structure there is in $f(x)$ separated by a distance of L. This can be understood most easily by considering the ACF of a function that consists of a few sharp peaks, such as that shown in Fig. 2.11. Basically, two spikes at x_1 and x_2 in $f(x)$, with amplitudes A_1 and A_2, will contribute a symmetric pair of very sharp components at $\pm\,(x_1-x_2)$, and magnitude $A_1 A_2$, towards the ACF of $f(x)$; they will also add an amount $A_1^2+A_2^2$ to the ACF at the origin.

$$\mathrm{ACF}(-x) \;=\; \mathrm{ACF}(x)^*$$

The reason for discussing the ACF is its linear relationship to the modulus of a Fourier transform:

$$\mathrm{ACF}(x) \;=\; \int\limits_{-\infty}^{\infty} \left|\mathrm{F}(k)\right|^2 \mathrm{e}^{\mathrm{i}kx}\,\mathrm{d}k \,, \qquad (2.56)$$

where $\mathrm{F}(k)$ is given by eqn (2.46). While a Fourier transform and its inverse contain the same information, albeit in different ways, and it's possible to switch between one and the other through eqns (2.45) and (2.46), the situation becomes less straightforward if only $\left|\mathrm{F}(k)\right|$ is available. We can begin to appreciate the problems caused by such a loss of the Fourier phase by comparing the relative complexity of the ACF with $f(x)$ in Fig. 2.11. The ACF, which is directly available

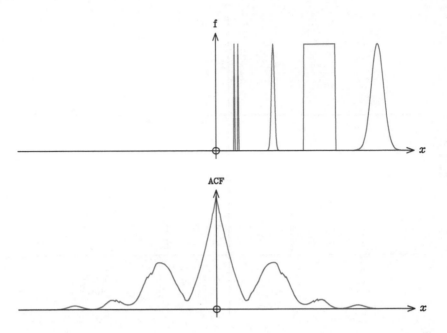

Fig. 2.12 An f(x) containing a variety of peaks and its auto-correlation function.

from $\left|F(k)\right|$ through eqn (2.56), is much harder to interpret in terms of the underlying structure; for a diffuse case, such as that in Fig. 2.12, it's almost impossible.

2.5   Fourier optics and physical insight

So far, we have discussed Fourier transforms in a largely abstract context. Now let's try to gain some physical insight into their properties with the aid of diffraction experiments familiar from high school physics. First, though, we need to establish the link between optics and Fourier transforms.

The geometry of the diffraction experiment is shown in Fig. 2.13, where a travelling plane wave passes through a set of slits and produces a pattern of dark and light bands on a very distant screen. We have made the problem one-dimensional for simplicity, but will indicate its generalization later. The nature of the aperture is defined by the function $A(x)$, which describes how much light passes through it at position x; this is called the *aperture function*. It usually only takes values of zero or one, corresponding to complete opaqueness and transparency respectively, but it could in principle be complex with $0 \leqslant \left|A(x)\right| \leqslant 1$.

To calculate the diffraction pattern, the principle of superposition tells us that we need to add up all the waves that emerge through the aperture. The amplitude of the contribution from the narrow region between x and $x + \Delta x$ is proportional to $A(x)\,\Delta x$, but what

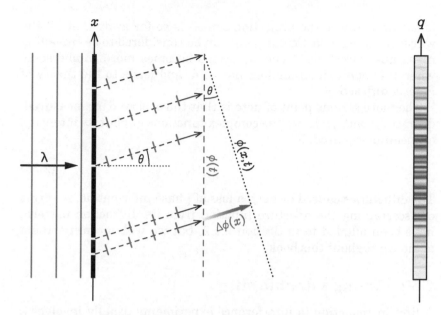

Fig. 2.13 The geometry for Fraunhofer diffraction by a one-dimensional aperture, $A(x)$; the interference pattern of interest, $I(q)$, is projected onto a distant screen.

about its phase ϕ? That depends on both x and the angle of propagation relative to the incident wave, θ, as well as the time t. The phase will be invariant with position parallel to the incoming wavefront, but will gain a relative factor of

$$\Delta\phi \;=\; \left(\tfrac{2\pi}{\lambda}\right) x \sin\theta$$

in the direction of θ due to the associated path difference of $x \sin\theta$. Hence, the complex contribution to the resultant wave is

$$\Delta\psi \;=\; \psi_\mathrm{o}\, A(x)\, e^{iqx}\, \Delta x \,,$$

where $q = 2\pi \sin\theta/\lambda$ and the temporal variation has been absorbed into the 'constant' of proportionality, ψ_o. The diffracted wave, ψ, is the sum of all such terms; in the limit $\Delta x \to 0$, it becomes the Fourier transform of the aperture function:

$$\psi(q) \;=\; \psi_\mathrm{o} \int\limits_{-\infty}^{\infty} A(x)\, e^{iqx}\, \mathrm{d}x \,. \tag{2.57}$$

Thus we met Fourier transforms a (very) long time ago but did not realize it! Before reminding ourselves of the results from elementary diffraction experiments, and trying to understand them in terms what we've now learnt about Fourier transforms, we need to make a few qualifying remarks.

The first point is essentially a technicality, but the above analysis assumes that we are considering *Fraunhofer* diffraction. This is the limit where the projection screen is so far away that all the waves reaching a particular point can be considered to be travelling in parallel directions. The equations becomes more cumbersome when this approximation does not hold, and leads to the theory of *Fresnel* diffraction.

The more serious point of note is that the observed, or measured, diffraction pattern is not the complex function $\psi(q)$ but its intensity, or modulus-squared, $I(q)$:

$$I(q) \;=\; \left| \psi(q) \right|^{2} \;=\; \psi(q)\,\psi(q)^{*}. \tag{2.58}$$

The difficulties caused by such a loss of phase information, in terms of ascertaining the aperture function from its diffraction pattern, have been alluded to in Section 2.4.2, but we will encounter them again throughout this book.

2.5.1 Young's double slit

A first introduction to interference experiments usually involves a *Young's double slit*. This consists of a pair of very narrow slits that are separated by a distance d, and give rise to a diffraction pattern of uniformly spaced dark and light bands which become closer together as d increases. Let's try to understand this theoretically by using eqns (2.57) and (2.58).

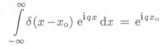

The aperture function for a Young's double slit can be modeled by two δ-functions located at a distance of $d/2$ on either side of an arbitrarily defined origin,

$$A(x) \;=\; \delta\!\left(x - \tfrac{d}{2}\right) + \delta\!\left(x + \tfrac{d}{2}\right),$$

$$\int_{-\infty}^{\infty} \delta(x - x_{\mathrm o})\,\mathrm{e}^{\mathrm{i}qx}\,\mathrm{d}x \;=\; \mathrm{e}^{\mathrm{i}qx_{\mathrm o}}$$

and is plotted in Fig. 2.14. Since δ-functions are easy to integrate, from eqn (2.54), the Fourier transform of $A(x)$ is readily shown to yield

$$\psi(q) \;=\; \psi_{\mathrm o}\left(\mathrm{e}^{\mathrm{i}qd/2} + \mathrm{e}^{-\mathrm{i}qd/2}\right)$$

$$=\; \psi_{\mathrm o}\,2\cos\!\left(\tfrac{qd}{2}\right),$$

where we have used eqn (2.32) in writing the second line. The product of this diffracted wave with its complex conjugate, $\psi(q)^{*}$, leads to the prediction

$$I(q) \;\propto\; \left[\cos\!\left(\tfrac{qd}{2}\right)\right]^{2} \;\propto\; 1 + \cos(qd)\,, \tag{2.59}$$

$$\cos 2\theta \;=\; 2\cos^{2}\theta - 1$$

where all the multiplicative prefactors not involving q, such as $\left|\psi_{\mathrm o}\right|^{2}$, have been omitted and a trigonometric double angle formula used on the far right-hand side. This pattern of 'uniform cosine fringes' is plotted in Fig. 2.14.

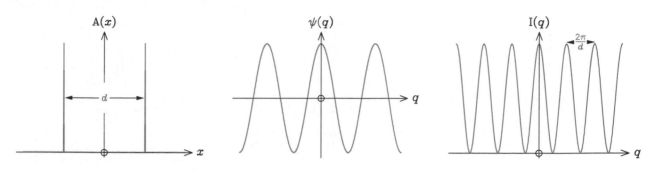

Fig. 2.14 The aperture function for a Young's double slit, $A(x)$, its Fourier transform, $\psi(q)$, and the diffraction pattern, $I(q)$.

The theoretical result in eqn (2.59) is consistent with the experimental observations: the dark and light bands are equally spaced, of uniform intensity and become closer together in inverse proportion to the distance d between the slits. The last feature is a universal property of Fourier transforms: the length scales which characterize a function and its Fourier transform are inversely related to each other. This leads to the use of the terminology *reciprocal space* when referring to the Fourier domain.

2.5.2 A single wide slit

Another common interference experiment involves a single wide slit that gives rise to a diffraction pattern where the intensity of the light bands diminishes rapidly away from a central bright region, which is itself twice as broad as the rest. Let's also try to understand this theoretically.

If we take the x-origin to be in the middle of the slit of width w, then the aperture function becomes

$$A(x) = \begin{cases} 1 & \text{if } |x| \leqslant \frac{w}{2}, \\ 0 & \text{otherwise}, \end{cases}$$

and is plotted in Fig. 2.15. According to eqn (2.57), therefore,

$$\psi(q) = \psi_\text{o} \int\limits_{-w/2}^{w/2} e^{iqx} \, dx \,.$$

This Fourier transform is easy to evaluate, because the integration of an exponential is straightforward, and yields

$$\frac{d}{dx}\left(e^{\mu x}\right) = \mu \, e^{\mu x}$$

$$\psi(q) = \psi_\text{o} \left[\frac{e^{iqx}}{iq} \right]_{-w/2}^{w/2} = \frac{\psi_\text{o}}{iq} \left(e^{iqw/2} - e^{-iqw/2} \right) .$$

The difference of the imaginary exponentials on the far right-hand side can be recognized as being equal to $2i$ times $\sin(qw/2)$ from

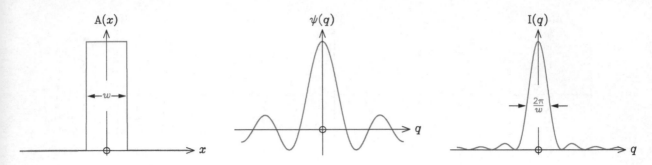

Fig. 2.15 The aperture function for a single wide slit, $A(x)$, its Fourier transform, $\psi(q)$, and the diffraction pattern, $I(q)$.

eqn (2.32). With this substitution, the modulus-squared of $\psi(q)$ leads to the prediction

$$I(q) \propto \left[\tfrac{1}{q}\sin\left(\tfrac{qw}{2}\right)\right]^2 \propto \frac{1-\cos(qw)}{q^2} \,, \qquad (2.60)$$

$$\operatorname{sinc}\theta = \frac{\sin\theta}{\theta} \to 1 \text{ as } \theta \to 0$$

which is shown in Fig. 2.15 and consistent with the *sinc*-squared behaviour of the observed diffraction pattern. We again see the inverse relationship between the width of the aperture function and the spread of the diffraction pattern: as one of them becomes broader the other gets narrower.

2.5.3   A diffraction grating

A diffraction grating is an aperture consisting of a large number of thin, parallel and equally spaced lines. In one dimension, it can be modelled as a periodic array of δ-functions:

$$A(x) = \sum_{m=-\infty}^{\infty} \delta(x-md) \,,$$

$$\int_{-\infty}^{\infty} \delta(x-md)\,e^{iqx}\,dx = e^{iqmd}$$

where d is the distance between the grating lines. Swapping the order of integration and summation, and using eqn (2.54), the Fourier transform of eqn (2.57) reduces to

$$\psi(q) = \psi_{\mathrm{o}} \sum_{m=-\infty}^{\infty} e^{iqdm} \,.$$

The nature of $\psi(q)$ becomes apparent once we realize that it's proportional to the sum of complex numbers that are of unit magnitude but varying phase. They will add up coherently if the product qd is an integer number of 2π, yielding a huge resultant sum, but cancel out otherwise. Hence,

$$\psi(q) \propto \sum_{n=-\infty}^{\infty} \delta(q-nq_{\mathrm{o}}) \,, \qquad (2.61)$$

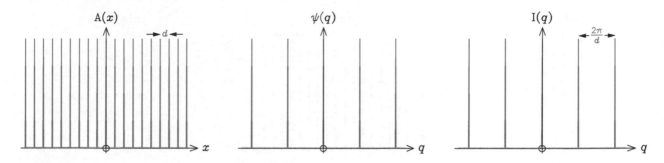

Fig. 2.16 The aperture function for a diffraction grating, $A(x)$, its Fourier transform, $\psi(q)$, and the diffraction pattern, $I(q)$.

where $q_0 = 2\pi/d$. The diffraction pattern has the same structure as the grating, therefore, but the spacing of the lines is inversely related to d (Fig. 2.16).

In terms of the physical set up of Fig. 2.13, where $q = 2\pi \sin\theta/\lambda$, sharp bright lines are seen when

$$n\lambda = d \sin\theta, \qquad (2.62)$$

$$\frac{2\pi \sin\theta}{\lambda} = \frac{2\pi n}{d}$$

for $n = 0, \pm1, \pm2, \ldots, \pm n_{\mathrm{max}}$, where the trigonometric constraint that $|\sin\theta| \leqslant 1$ imposes a cutoff on the highest observable order n_{max}. If the spacing of the diffraction grating is known, such as 500 lines per millimetre (so that $d = 2\,\mu\mathrm{m}$), then eqn (2.62) provides the basis for an accurate measurement of the wavelength of the illumination. If white light is used for the experiment, then the intense central line is accompanied by increasingly dispersed rainbows for the higher orders; this is because each of the wavelengths that makes up white light satisfies eqn (2.62) for a slightly different angle θ for a given value of $n \neq 0$.

2.5.4 The convolution theorem in action

Although a real diffraction grating isn't infinite as assumed above, we expect the analysis to be a very good approximation for one that is sufficiently large. The case of a grating of limited extent w can be addressed by combining the results of eqns (2.60) and (2.61) through the convolution theorem: as the aperture function can be expressed as a product of an infinite grating with line spacing d and a single slit of width w, as illustrated in Fig. 2.17, the Fourier transform of the finite grating is equal to the convolution of the Fourier transforms of the infinite grating and the single wide slit. The resultant diffraction pattern is simply that of the infinite grating but with each of the δ-functions replaced by a narrow sinc-squared function, as shown in Fig. 2.17. A qualification is in order here, in that $I(q) \propto |G(q)|^2 \otimes |H(q)|^2$ is only an approximation (albeit a good one); strictly speaking, $I(q) \propto |G(q) \otimes H(q)|^2$. Given the inverse relationship between the length scales of a function and its Fourier trans-

$$A(x) = g(x) \times h(x)$$

$$\therefore \ \psi(q) = \psi_0\, G(q) \otimes H(q)$$

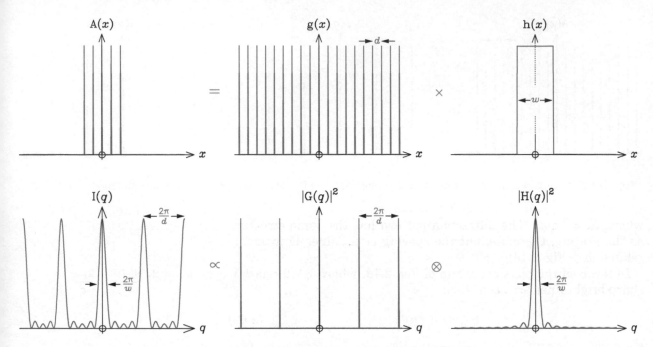

Fig. 2.17 The diffraction pattern from a grating of limited extent, w, can be evaluated from a knowledge of the Fourier transforms of an infinite grating, with line spacing d, and a single slit, of width w, through the use of the convolution theorem.

form, the width of the large diffraction peaks tells us about the size w of the grating whereas the distance between them indicates the d-spacing of its lines. As the number of grating lines goes up, so that the ratio w/d increases, the principal peaks become narrower and more low-level wiggles appear between them.

The convolution theorem also enables us to ascertain the diffraction pattern for a pair of broad slits from the results of eqns (2.59) and (2.60). Taking each to be of width w, and separated by d, the aperture function can be seen as a convolution of an ideal Young's double slit with a narrow but finite single slit, as in Fig. 2.18. Since the Fourier transform of the former is then equal to the product of those of the latter, the intensity of the uniform cosine fringes that we'd expect from a perfect Young's double slit is modulated by a slowly varying sinc-squared function.

2.5.5 Multi-dimensional generalization

$$A(x) = g(x) \otimes h(x)$$

$$\therefore \ \psi(q) = \psi_\circ \, G(q) \times H(q)$$

Having illustrated Fourier transforms and the use of the convolution theorem with one-dimensional versions of familiar high school experiments, let's indicate the multi-dimensional generalization of eqn (2.57). A closer examination of Fig. 2.13 reveals q to be the x-component of the wavevector **k** of Section 2.1.1:

$$\mathbf{k} = (k_x, k_y, k_z)$$

$$k_x = \left(\tfrac{2\pi}{\lambda}\right)\sin\theta = q \quad \text{and} \quad k_z = \left(\tfrac{2\pi}{\lambda}\right)\cos\theta,$$

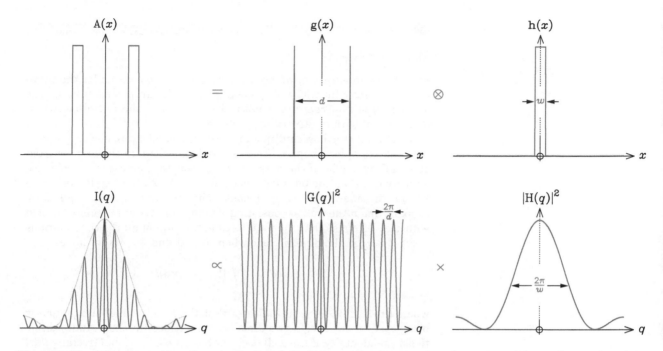

Fig. 2.18 The diffraction pattern from a pair of slits of width w and separation d can be evaluated from a knowledge of the Fourier transforms of a Young's double slit, of spacing d, and a single slit, of width w, with the convolution theorem.

where we have taken z to be the original direction of propagation, from the aperture to the projection screen, and $|\mathbf{k}| = 2\pi/\lambda$. With this observation, it seems plausible that the two-dimensional diffraction pattern, $I(k_x, k_y)$, from an aperture in the x–y plane, $A(x, y)$, with y coming out of the page in Fig. 2.13, might be given by the modulus-squared of

$$\psi(k_x, k_y) = \psi_\circ \int\limits_{-\infty}^{\infty} \int\limits_{-\infty}^{\infty} A(x, y)\, e^{i(k_x x + k_y y)}\, dx\, dy \,. \qquad (2.63)$$

This *double integral*, over the surface area of the aperture, simplifies to the product of two one-dimensional integrals if the aperture function is *separable*:

$$\psi(k_x, k_y) = \psi_\circ \int\limits_{-\infty}^{\infty} A_1(x)\, e^{i k_x x}\, dx \int\limits_{-\infty}^{\infty} A_2(y)\, e^{i k_y y}\, dy$$

if $A(x, y) = A_1(x)\, A_2(y)$. The substitution of either $\delta(y)$ or a constant for $A_2(y)$, and the properties of δ-functions, allows us to confirm that eqn (2.63) reduces to the one-dimensional form of eqn (2.57) if the aperture is either infinitesimally thin or invariant with respect to y. Strictly speaking,

$$\int\limits_{-\infty}^{\infty} e^{i(q-q_\circ)t}\, dt = 2\pi\, \delta(q-q_\circ)$$

$$\psi \to \psi_1(k_x) \quad \text{as} \quad A \to A_1(x)\, \delta(y) \quad \text{and} \quad \psi \to \psi_1(k_x)\, \delta(k_y) \quad \text{as} \quad A \to A_1(x)$$

From Sivia and Rawlings (1999),
Foundations of Science Mathematics,
Oxford Chemistry Primers Series, **77**.

Multiple integrals

In ordinary integration, we are concerned with the area under the curve $y = f(x)$. Many functions of interest in real life entail several variables, and *multiple integrals* are a natural extension of the one-dimensional ideas to deal with multivariate problems.

To get a feel for how multiple integrals arise, let's consider a couple of physical examples. Suppose that we wish to calculate the force exerted on a wall by a gale. If the pressure P was constant across the whole face with area A, then the total force is simply $P \times A$. With a varying pressure $P(x, y)$, the answer is not so obvious. This situation can be handled by thinking about the wall as consisting of many small square segments, each with area $\delta x \, \delta y$, so that the total force is the sum of all the contributions $P(x, y) \, \delta x \, \delta y$; in the limiting case when $\delta x \to 0$ and $\delta y \to 0$, we have

$$\text{Force} = \iint_{\text{wall}} P(x, y) \, dx \, dy$$

where the double integral indicates that the infinitesimal summation is being carried out over a two-dimensional surface (in the x and y directions). Incidentally, if the wall does not have a conventional (rectangular) shape then its area can be calculated similarly according to

$$\text{Area} = \iint_{\text{wall}} dx \, dy \, .$$

The double integral is also called a *surface integral*.

Another illustration is provided by quantum mechanics where the modulus-squared of the wave function, $|\psi(x, y, z)|^2$, of an electron (say) gives the *probability density* of finding it at some point in space. The chances that the electron is in a small (cuboid) region of volume $\delta x \, \delta y \, \delta z$ is then $|\psi(x, y, z)|^2 \, \delta x \, \delta y \, \delta z$. Hence, the probability of finding it within a finite domain V is given by

$$\text{Probability} = \iiint_{\text{V}} |\psi(x, y, z)|^2 \, dx \, dy \, dz \, ,$$

which is known as a *triple*, or *volume*, *integral*.

and demonstrates the reciprocal Fourier relationship between the widths of $A_2(y)$ and $\psi_2(k_y)$ in the limit of complete invariance versus a δ-function. A careful consideration of the situation, in a manner analogous to that used to derive eqn (2.57), shows eqn (2.63) to be the correct two-dimensional extension.

The most common case of two-dimensional diffraction is from a circular hole, but a rectangle is easier to deal with analytically. This is because the aperture function of the latter, which is equal to one inside the rectangle and zero outside it, is separable and yields a

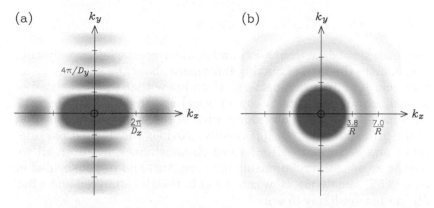

Fig. 2.19 Diffraction patterns from two-dimensional apertures: (a) a rectangular opening of size D_x by D_y, and (b) a circular hole of radius R.

Fourier transform that is just a product of the now familiar sinc functions in k_x and k_y; its modulus-squared is shown in Fig. 2.19(a). The evaluation of the integral of eqn (2.63) is less straightforward for a circular aperture, but the resultant diffraction pattern is plotted in Fig. 2.19(b). It is circularly symmetric, depending only on $k_x^2 + k_y^2$, and is similar to a sinc function in the radial direction; the behaviour is formally governed by a J_1 *Bessel function*. The central bright region is called an *Airy disc*, and its spread is the basis of the resolution formula of eqn (1.8).

Having seen the formulae for the Fourier transforms of one- and two-dimensional functions, in eqns (2.45), (2.46) and (2.63), we can state the M-dimensional generalization succinctly by using vector notation:

$$\mathrm{f}(\mathbf{r}) = (2\pi)^{-\frac{M}{2}} \int_{-\infty}^{\infty} \int_{-\infty}^{\infty} \cdots \int_{-\infty}^{\infty} \mathrm{F}(\mathbf{k})\, \mathrm{e}^{-\mathrm{i}\mathbf{k}\bullet\mathbf{r}}\, \mathrm{d}^M \mathbf{k} \qquad (2.64)$$

and

$$\mathrm{F}(\mathbf{k}) = (2\pi)^{-\frac{M}{2}} \int_{-\infty}^{\infty} \int_{-\infty}^{\infty} \cdots \int_{-\infty}^{\infty} \mathrm{f}(\mathbf{r})\, \mathrm{e}^{\mathrm{i}\mathbf{k}\bullet\mathbf{r}}\, \mathrm{d}^M \mathbf{r} \qquad (2.65)$$

$$\mathrm{d}^3\mathbf{k} = \mathrm{d}k_x\, \mathrm{d}k_y\, \mathrm{d}k_z$$

$$\mathbf{k}\bullet\mathbf{r} = k_x x + k_y y + k_z z + \cdots$$

$$\mathrm{d}^3\mathbf{r} = \mathrm{d}x\, \mathrm{d}y\, \mathrm{d}z$$

where $\mathbf{r} = (x, y, z, \dots)$ and $\mathbf{k} = (k_x, k_y, k_z, \dots)$ have M components, with corresponding hyper-volume elements $\mathrm{d}^M \mathbf{r}$ and $\mathrm{d}^M \mathbf{k}$. We reiterate that a Fourier transform and its inverse come as a linked pair, but which one is called which is arbitrary. Their precise definitions are also a matter of convention. No multiplicative prefactors are required if the wavevector is specified in cycles rather than radians per unit length, for example, when the $\mathbf{k}$ in the exponents is replaced with $2\pi\mathbf{k}$.

2.6   Fourier data analysis

The analysis of data from X-ray and neutron scattering experiments is similar to the task of making inferences about the aperture function from its diffraction pattern. If we knew that $A(x)$ consisted of a small number of slits, n say, of equal spacing d, as in Fig. 2.17 with $w = (n-1)d$, then an examination of the width and separation of the principal peaks in $I(q)$ readily provides the desired parameters n and d. In less well informed circumstances, however, all we have to go on is the relationship between $A(x)$ and $I(q)$ enshrined in eqns (2.57) and (2.58). How can the data then be analysed and what difficulties are likely to arise?

2.6.1   The phase problem

Ignoring matters of practicality for the moment, the most relevant mathematical operation that can be performed on a diffraction pattern is a Fourier transform:

$$\int_{-\infty}^{\infty}\int_{-\infty}^{\infty}\cdots\int_{-\infty}^{\infty} I(\mathbf{k})\, e^{-i\mathbf{k}\bullet\mathbf{r}}\, d^M\mathbf{k} \;\propto\; \int_{-\infty}^{\infty}\int_{-\infty}^{\infty}\cdots\int_{-\infty}^{\infty} A(\mathbf{t})^* \, A(\mathbf{r}+\mathbf{t})\, d^M\mathbf{t}\,,$$

giving the M-dimensional generalization of the ACF of eqns (2.55) and (2.56), where

$$I(\mathbf{k}) = \big|\psi(\mathbf{k})\big|^2 \quad\text{and}\quad \psi(\mathbf{k}) = \psi_\circ \int_{-\infty}^{\infty}\int_{-\infty}^{\infty}\cdots\int_{-\infty}^{\infty} A(\mathbf{r})\, e^{i\mathbf{k}\bullet\mathbf{r}}\, d^M\mathbf{r}\,.$$

Whereas the correspondence between $A(\mathbf{r})$ and the complex function $\psi(\mathbf{k})$ is one-to-one, implying that there is no loss of information in the transformation, the same is not true of $A(\mathbf{r})$ and the real and positive diffraction pattern $I(\mathbf{k})$. Only the auto-correlation function of $A(\mathbf{r})$ can be ascertained unambiguously from $I(\mathbf{k})$, and we have already seen, in Figs. 2.11 and 2.12, how much more difficult it is to interpret the ACF than $A(\mathbf{r})$.

The simplest way of appreciating how the lack of phase, $\arg\{\psi(\mathbf{k})\}$, in a diffraction pattern results in a loss of uniqueness about $A(\mathbf{r})$ is to consider the Fourier transform of an aperture function that has been shifted by $\mathbf{r}_\circ$,

$$\psi_\circ \int_{-\infty}^{\infty}\int_{-\infty}^{\infty}\cdots\int_{-\infty}^{\infty} A(\mathbf{r}+\mathbf{r}_\circ)\, e^{i\mathbf{k}\bullet\mathbf{r}}\, d^M\mathbf{r} \;=\; \psi(\mathbf{k})\, e^{-i\mathbf{k}\bullet\mathbf{r}_\circ}\,,$$

which differs from that of $A(\mathbf{r})$ only through an additional factor of $-\mathbf{k}\bullet\mathbf{r}_\circ$ in its argument; the intensity, or modulus-squared,

$$e^\theta e^{-\theta} = e^0 = 1 \qquad\qquad \big[\psi(\mathbf{k})\, e^{-i\mathbf{k}\bullet\mathbf{r}_\circ}\big]\big[\psi(\mathbf{k})\, e^{-i\mathbf{k}\bullet\mathbf{r}_\circ}\big]^* \;=\; \psi(\mathbf{k})\,\psi(\mathbf{k})^*\, e^{-i\mathbf{k}\bullet\mathbf{r}_\circ}\, e^{i\mathbf{k}\bullet\mathbf{r}_\circ}\,,$$

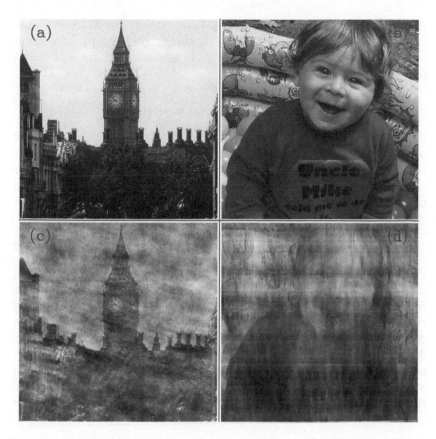

Fig. 2.20 The phase problem: (c) has the Fourier phases of (a) and the Fourier amplitudes of (b), while (d) has the phases of (b) and the amplitudes of (a).

is unchanged. The amplitude of a Fourier transform is, therefore, insensitive to translation. The same is true of the inversion of a real function, so that $A(\mathbf{r})$ and $A(-\mathbf{r})$ give identical diffraction patterns if $A(\mathbf{r}) = A(\mathbf{r})^*$. This provides another elementary demonstration of the loss of uniqueness without the Fourier phase.

The importance of the phase of a Fourier transform can be illustrated dramatically with graphical examples of the type shown in Fig. 2.20. Here two photographs, pertaining to any subject or scene, are Fourier transformed numerically and the phases of one assigned to the amplitudes of the other. Each of these hybrid sets of complex coefficients is then inverse Fourier transformed, and the resultant pictures examined visually. Instinctively we would guess that the outcome of this numerical experiment will be a complete mess, for why should the Fourier phases of one distribution of light intensity have anything to do with the amplitudes from another; if not, we might expect to see some sort of mixture of the two sources. What we find in practice is certainly degraded compared to the originals, but each output only resembles the scene which contributed the Fourier phase with no hint of that from which the amplitudes were taken.

It seems that most of the structural information in a Fourier transform resides in its phase; and since this is missing in diffraction data, it makes their analysis difficult in general without additional prior knowledge.

2.6.2 Truncation effects and windowing

Even when the main interest is in the ACF of the aperture function, and the absence of Fourier phase is not a problem, the limited sampling of a diffraction pattern causes difficulties in practice. In the simplest one-dimensional case, when $I(q)$ is available only within the range $|q| \leqslant q_{max}$, the truncated Fourier integral

$$\int_{-q_{max}}^{q_{max}} I(q)\, e^{-iqx}\, dq \;=\; 2 \int_{0}^{q_{max}} I(q)\, \cos(qx)\, dq\,, \qquad (2.66)$$

$$A(x) = A(x)^* \;\Longrightarrow\; I(q) = I(-q)$$

where the cosine equivalent on the right assumes that the aperture function is real, yields an estimate of the ACF that is corrupted by ripples with a characteristic wavelength of $2\pi/q_{max}$. These artefacts can be understood with the aid of the convolution theorem, and Fig. 2.21, by considering eqn (2.66) to be the Fourier transform of the product of the full but unmeasured diffraction pattern, $J(q)$, and a 'top-hat' function of width $2\,q_{max}$, $H(q)$. The result is, therefore, the true but unknown auto-correlation function, acf, convolved with a

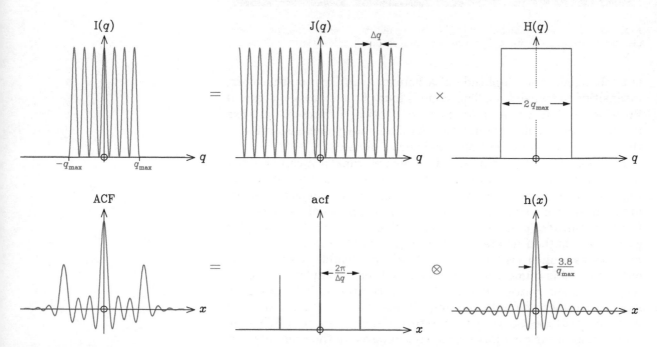

Fig. 2.21 The Fourier transform of a diffraction pattern of limited q-extent, $I(q)$, yields an ACF of the aperture function which is corrupted by truncation ripples associated with q_{max}; their origin is easily understood from the convolution theorem.

sinc function, $h(x)$, whose central peak has a full width at half maximum (FWHM) of about $3.8/q_{max}$.

The messy picture due to the truncation ripples can be cleaned up greatly by multiplying the incomplete diffraction pattern, $I(q)$, with a *window* function, $W(q)$, which decays smoothly from one at the origin to zero around $\pm q_{max}$, before the (inverse) Fourier transform is calculated. This is illustrated in Fig. 2.22 with the ubiquitous *Gaussian*,

$$W(q) = \exp\left(-\frac{q^2}{2\,\sigma^2}\right), \qquad (2.67)$$

$$\text{FWHM} = \sqrt{8\ln 2}\ \sigma \approx 2.35\,\sigma$$

whose *standard deviation* σ was chosen somewhat arbitrarily as $q_{max}/2$. The resultant auto-correlation function, denoted by acf, is said to be a *filtered* version of the ACF given by $I(q)$. The suppression of the truncation ripples can also be understood from the convolution theorem, which tells us that $\text{acf}(x) = \text{ACF}(x) \otimes w(x)$, because the subsidiary oscillations are averaged out through a blurring with the filter $w(x)$. The latter is just the Fourier transform of the windowing function, $W(q)$, with

$$w(x) \propto \exp\left(-\frac{\sigma^2 x^2}{2}\right) \qquad (2.68)$$

for the case of eqn (2.67). Although the spurious peaks and troughs are increasingly reduced as $w(x)$ becomes broader, requiring $W(q)$

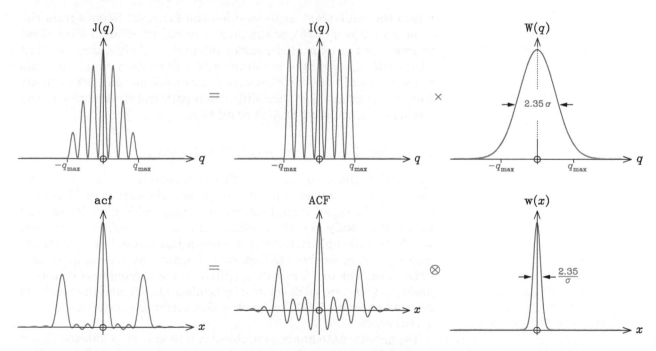

Fig. 2.22 Truncation ripples can be suppressed by multiplying the diffraction pattern, $I(q)$, by a 'window' function, $W(q)$, which decays smoothly to zero over a q-range comparable to that of the measurements, before calculating the (inverse) Fourier transform; the resultant 'filtered' auto-correlation function can also be understood from the convolution theorem.

to be narrower, the drawback is that intrinsically sharp features of the auto-correlation function are smeared out even more. Thus filtering is a matter of striking a balance between the suppression of the truncation ripples and a further loss of resolution. A variety of windowing functions have been developed to try to best achieve this end.

Diffraction measurements are often unattainable at low q-values as well as high ones, so that $I(q)$ is available only within the range $q_{\min} \leqslant |q| \leqslant q_{\max}$. The truncated Fourier integral,

$$\int_{q_{\min}}^{q_{\max}} I(q) \, \cos(qx) \, dq \,,$$

then yields an estimate of the ACF of the aperture function that is plagued by both low and high frequency artefacts. Structure in $A(x)$ which is longer than around $2\pi/q_{\min}$ or narrower than about $2\pi/q_{\max}$ cannot be inferred reliably. The difficulty caused by a lack of $I(0)$ is easiest to appreciate since, from eqns (2.57) and (2.58), it relates to the area under $A(x)$:

$$I(0) \propto \left| \int_{-\infty}^{\infty} A(x) \, dx \right|^2 \propto \int_{-\infty}^{\infty} ACF(x) \, dx \,, \qquad (2.69)$$

where the equivalent expression on the far right follows from the x-integral of eqn (2.55), or the inverse of eqn (2.56) with k (or q) set to zero. As the truncated Fourier integral implicitly assumes that $I(0) = 0$ if $q_{\min} \neq 0$, the resultant ACF will contain equal amounts of positive and negative structure to ensure a net null area. Apart from at the origin, $q = 0$, the diffraction pattern is insensitive to the addition of a constant to $A(x)$ or its ACF.

2.6.3 Noise and probability theory

In practice, the analysis of a diffraction pattern is also limited by the *noise* in the measurement process and the extent to which the details of the experimental setup are understood and modelled. The task is not really one of calculating an inverse Fourier transform, which isn't possible in a strict mathematical sense, but a matter of making *inferences* about the aperture function given incomplete and noisy data. The tool for dealing with and quantifying uncertainty is *probability theory*, as developed by Laplace (1812), and the reader is referred to Sivia (1996) for an extended tutorial. A brief overview is given below.

Data analysis: a Bayesian tutorial, Sivia (1996), Oxford University Press; 2nd edition (2006) with Skilling.

The generic data analysis problem can be stated as follows: Given a set of N measurements $\{D_k\}$, for $k = 1, 2, 3, \ldots, N$, and some pertinent information H, what can we infer about the object of interest $A(x)$? The Fourier nature of the experiment enters the analysis through the equation that predicts the kth data point, F_k, for a given aperture $\mathrm{A}(x)$:

$$F_k = \mathrm{f}\!\left(\mathrm{I}(q), k\right),\tag{2.70}$$

where

$$\mathrm{I}(q) = \left|\, \psi_\circ \int_{-\infty}^{\infty} \mathrm{A}(x)\,\mathrm{e}^{\mathrm{i}qx}\,\mathrm{d}x \right|^{2}\tag{2.71}$$

and 'f' is the function that models the measurement process. In the simplest case $F_k = \mathrm{I}(q_k)$, but a more common situation involves the convolution of $\mathrm{I}(q)$ with an instrumental *response*, or *resolution*, function, $\mathrm{R}(q)$, and the addition of a slowly varying background signal, $\mathrm{B}(q)$:

$$F_k = \int_{-\infty}^{\infty} \mathrm{I}(q)\,\mathrm{R}(q_k\!-\!q)\,\mathrm{d}q\; +\mathrm{B}(q_k)\,.\tag{2.72}$$

The noise, or the expected mismatch between F_k and D_k, is usually quantified through an *error-bar*, σ_k, which is a shorthand way of assigning a Gaussian probability for the likelihood of the kth datum:

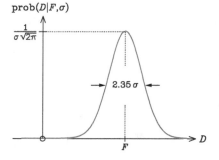

$$\mathrm{prob}\!\left(D_k \,\middle|\, \mathrm{A}(x), H\right) = \frac{1}{\sigma_k\sqrt{2\pi}}\,\exp\!\left[-\frac{(D_k\!-\!F_k)^2}{2\,\sigma_k^2}\right],\tag{2.73}$$

where the vertical bar '|' means 'given' (so that all items to the right of this conditioning symbol are taken as being true) and the comma is read as the conjunction 'and'. A knowledge of eqns (2.70)–(2.72) and, hopefully, the related resolution and background functions, as well as the error-bars, is implicitly assumed in H. If the N measurements, $\{D_k\}$, are *independent*, in that the noise associated with one is unrelated to that of another (as far as H is concerned), then their joint likelihood is just the product of the individual contributions:

$$\mathrm{prob}\!\left(\{D_k\} \,\middle|\, \mathrm{A}(x), H\right) = \prod_{k=1}^{N} \mathrm{prob}\!\left(D_k \,\middle|\, \mathrm{A}(x), H\right).$$

In conjunction with eqn (2.73), therefore, the *likelihood function* for the data can be written as

$$\mathrm{prob}\!\left(\{D_k\} \,\middle|\, \mathrm{A}(x), H\right) \propto \exp\!\left(-\frac{\chi^2}{2}\right),\tag{2.74}$$

where

$$\chi^2 = \sum_{k=1}^{N}\left(\frac{F_k\!-\!D_k}{\sigma_k}\right)^{2}\tag{2.75}$$

is the sum of the squares of the *normalized residuals*.

Our inference, or 'state of knowledge', about the aperture function in the light of the data and H is not encapsulated by the likelihood function but by the *posterior probability*,

$$\text{prob}\Big(\text{A}(x)\,\Big|\,\{D_k\},H\Big)\,,$$

where the positions of $\{D_k\}$ and $\text{A}(x)$ are reversed with respect to the conditioning symbol. The $\text{A}(x)$ which gives the largest value for the posterior probability can be regarded as the 'best' estimate of the aperture function, while the range of the alternatives that yield a reasonable fraction of the maximum probability gives an indication of the uncertainty. The likelihood function is related to the posterior probability through *Bayes' theorem*,

$$\text{prob}\Big(\text{A}(x)\,\Big|\,\{D_k\},H\Big) = \frac{\text{prob}\Big(\{D_k\}\,\Big|\,\text{A}(x),H\Big) \times \text{prob}\Big(\text{A}(x)\,\Big|\,H\Big)}{\text{prob}\Big(\{D_k\}\,\Big|\,H\Big)}\,,$$

where the second term in the numerator is called the *prior probability*, and represents our state of knowledge (or ignorance) about the aperture function before the analysis of the data, and the denominator usually constitutes an uninteresting proportionality constant (required for normalization) since it doesn't explicitly mention $\text{A}(x)$. The latter plays a crucial role when comparing different assumptions or models, however, such as H_1 versus H_2, and is referred to as the 'global likelihood', 'prior predictive' or simply the *evidence* in that context.

A quantitative discussion of the aperture function is contingent on a parametric description of $\text{A}(x)$, of course, and its choice is a reflection of the information H at hand. If it were known that we were dealing with a pair of slits of equal finite width, as in Fig. 2.18 for example, then $\text{A}(x)$ would be defined by the two parameters d and w as follows:

$$\text{A}(x) = \begin{cases} 1 & \text{if } \left|x \pm \frac{d}{2}\right| \leqslant \frac{w}{2}\,, \\ 0 & \text{otherwise}\,, \end{cases}$$

where $d > w > 0$. If very little information was available, then we might use the formulation

$$\text{A}(x) = \sum_{j=1}^{M} c_j\, \text{G}_j(x)\,,$$

where the M coefficients, $\{c_j\}$, define the aperture function through a linear combination of suitable *basis functions*, $\text{G}_j(x)$. Although a larger value of M provides greater flexibility in the range of $\text{A}(x)$ that can be modelled, a more careful choice of the $\text{G}_j(x)$ can reduce the number required and, thereby, aid many aspects of the data analysis task.

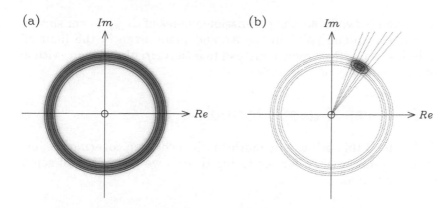

Fig. 2.23 The likelihood constraint on the value of a Fourier coefficient given (a) a noisy measurement of its modulus-squared and (b) additional phase information.

When the aperture function is defined by only a few parameters, the data tend to impose a strong constraint on their allowed values. The likelihood function dominates the posterior probability in this case and the prior, which is relatively broad to represent ignorance, is largely irrelevant. For the likelihood function of eqns (2.74) and (2.75), therefore, the optimal parameters of $A(x)$ are those that yield the smallest value of χ^2; this is called the *least-squares* estimate. When little is known about the aperture function beforehand, and its description entails a large number of parameters to reflect this initial ignorance, it becomes important to give due consideration to the prior to encode whatever weak information is available about $A(x)$. For example, positivity, bounds, local smoothness and so on. This leads to the use of *regularization* procedures, or constrained optimization, such as *maximum entropy*.

The computational task of finding the maximum of the posterior probability distribution and determining its spread, in the space of the parameters used to describe $A(x)$, can be a very challenging one. If we have a good initial estimate of the optimal solution, then an efficient gradient algorithm, such as *Newton–Raphson*, can often be employed. Otherwise we may need to use the slower, but more robust, *Monte Carlo* methods. These sorts of practical considerations can make it tempting to ignore the noise and limited coverage of the data, and try to emulate an inverse Fourier transform in some way. For the ACF, and with appropriate filtering, this can provide a useful quick method for a qualitative analysis.

As mentioned earlier, the loss of the Fourier phase in diffraction experiments causes a serious difficulty for ascertaining the aperture function. We can begin to appreciate this from a probabilistic point of view by considering the constraint that the likelihood function imposes on the value of a Fourier coefficient, $\psi(q_k)$, when only its modulus-squared can be measured; this is shown pictorially in Fig. 2.23(a). Unlike the case of Fig. 2.23(b), where additional phase information is available, the permissible values of $\psi(q_k)$ do not shrink towards a unique point in the Argand plane even in the limit of noiseless data; they reduce instead to a thin circular region, with a phase ambiguity of 2π radians.

2.7 A list of useful formulae

To finish off this principally mathematical chapter, covering the important prerequisites for scattering theory, we give a list of some useful formulae.

Powers and logarithms *(see Section 1.1)*

$$a^M a^N = a^{M+N}, \qquad (a^M)^N = a^{MN},$$

$$a^0 = 1, \qquad a^{-N} = 1/a^N \quad \text{and} \quad a^{1/p} = \sqrt[p]{a} \ \text{(integer p)}.$$

$$y = a^x \iff x = \log_a(y)$$

$$\log(AB) = \log(A) + \log(B) \qquad \text{and} \qquad \log(A/B) = \log(A) - \log(B),$$

$$\log(A^\beta) = \beta \log(A) \qquad \text{and} \qquad \log_b(A) = \log_a(A) \times \log_b(a).$$

Trigonometry

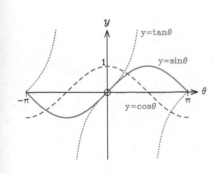

$$\sin\theta = \frac{y}{r} = \frac{1}{\operatorname{cosec}\theta}, \quad \cos\theta = \frac{x}{r} = \frac{1}{\sec\theta}, \quad \tan\theta = \frac{y}{x} = \frac{\sin\theta}{\cos\theta} = \frac{1}{\cot\theta}.$$

$$x^2 + y^2 = r^2 \iff \sin^2\theta + \cos^2\theta \equiv 1$$

$$\tan^2\theta + 1 \equiv \sec^2\theta$$

$$\cot^2\theta + 1 \equiv \operatorname{cosec}^2\theta$$

$$\sin(A \pm B) = \sin A \cos B \pm \cos A \sin B \implies \sin 2\theta = 2\sin\theta\cos\theta$$

$$\cos(A \pm B) = \cos A \cos B \mp \sin A \sin B \implies \cos 2\theta = \cos^2\theta - \sin^2\theta$$

$$2\sin A \cos B = \sin(A+B) + \sin(A-B)$$

$$2\cos A \cos B = \cos(A+B) + \cos(A-B)$$

$$-2\sin A \sin B = \cos(A+B) - \cos(A-B)$$

$$\frac{a}{\sin A} = \frac{b}{\sin B} = \frac{c}{\sin C}$$

$$c^2 = a^2 + b^2 - 2ab\cos C$$

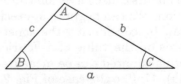

Power series, sums and expansions *(see Section 2.2.4)*

$$\sin x = x - \frac{x^3}{3!} + \frac{x^5}{5!} - \frac{x^7}{7!} + \cdots$$

$$\cos x = 1 - \frac{x^2}{2!} + \frac{x^4}{4!} - \frac{x^6}{6!} + \cdots$$

$$n! = n \times (n-1) \times (n-2) \times \cdots \times 3 \times 2 \times 1$$

$$e^x = \exp(x) = 1 + x + \frac{x^2}{2!} + \frac{x^3}{3!} + \frac{x^4}{4!} + \cdots$$

$$e = e^1 = 1 + 1 + \frac{1}{2} + \frac{1}{6} + \frac{1}{24} + \cdots$$
$$= 2.718\ldots$$

$$\log_e(1+x) = \ln(1+x) = x - \frac{x^2}{2} + \frac{x^3}{3} - \frac{x^4}{4} + \frac{x^5}{5} - \cdots \qquad \left(|x| < 1\right)$$

$$(1+x)^p = 1 + px + \frac{p(p-1)}{2!}x^2 + \frac{p(p-1)(p-2)}{3!}x^3 + \cdots \qquad \left(|x|<1\right) \qquad \sqrt{1+x} = 1 + \frac{x}{2} - \frac{x^2}{8} + \cdots$$

$$(a+b)^n = \sum_{k=0}^{n} {}^nC_k \, a^k \, b^{n-k}$$

$${}^nC_k = \binom{n}{k} = \frac{n!}{k!\,(n-k)!}, \qquad 0! = 1$$

$$\sum_{k=1}^{n} a + (k-1)d = \frac{n}{2}\left[2a + (n-1)d\right]$$

$$1 + 2 + 3 + \cdots + N = \frac{N(N+1)}{2}$$

$$\sum_{k=1}^{n} a\,r^{k-1} = \frac{a(1-r^n)}{1-r} \longrightarrow \frac{a}{1-r} \quad \text{as } n \to \infty \text{ and } |r| < 1$$

Vectors *(see Section 2.1.1)*

$$\mathbf{a} \times (\mathbf{b} \times \mathbf{c}) = (\mathbf{a} \cdot \mathbf{c})\,\mathbf{b} - (\mathbf{a} \cdot \mathbf{b})\,\mathbf{c}$$

Complex numbers *(see Section 2.2)*

$$\sinh x = \frac{e^x - e^{-x}}{2} = -i\sin(ix)$$

$$\cosh x = \frac{e^x + e^{-x}}{2} = \cos(ix)$$

$$\tanh x = \frac{\sinh x}{\cosh x} = \frac{e^{2x} - 1}{e^{2x} + 1}$$

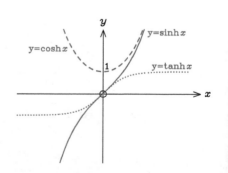

Differentiation and integration *(see Sections 2.1.2 and 2.3.1)*

$$y'' = \frac{\mathrm{d}^2 y}{\mathrm{d}x^2} = \frac{\mathrm{d}}{\mathrm{d}x}\left(\frac{\mathrm{d}y}{\mathrm{d}x}\right) \qquad \frac{\mathrm{d}^n y}{\mathrm{d}x^n} = \frac{\mathrm{d}}{\mathrm{d}x}\left(\frac{\mathrm{d}^{n-1} y}{\mathrm{d}x^{n-1}}\right), \qquad y' = \frac{\mathrm{d}y}{\mathrm{d}x} = \frac{1}{\mathrm{d}x/\mathrm{d}y},$$

$$\frac{\mathrm{d}}{\mathrm{d}x}(uv) = u\frac{\mathrm{d}v}{\mathrm{d}x} + v\frac{\mathrm{d}u}{\mathrm{d}x}, \qquad \frac{\mathrm{d}}{\mathrm{d}x}\left(\frac{u}{v}\right) = \frac{v\,u' - u\,v'}{v^2},$$

$$(uv)'' = u\,v'' + 2\,u'v' + u''v \qquad \frac{\mathrm{d}^n}{\mathrm{d}x^n}(uv) = \sum_{k=0}^{n} {}^n C_k \frac{\mathrm{d}^k u}{\mathrm{d}x^k}\frac{\mathrm{d}^{n-k} v}{\mathrm{d}x^{n-k}}, \qquad \frac{\mathrm{d}y}{\mathrm{d}x} = \frac{\mathrm{d}y}{\mathrm{d}u}\times\frac{\mathrm{d}u}{\mathrm{d}x} = \frac{\mathrm{d}y/\mathrm{d}t}{\mathrm{d}x/\mathrm{d}t}.$$

$$\boxed{\frac{\mathrm{d}f}{\mathrm{d}x} = g(x) \iff \int_a^b g(x)\,\mathrm{d}x = \Big[f(x)+c\Big]_a^b = f(b) - f(a)}$$

$$\frac{\mathrm{d}}{\mathrm{d}t}\int_{a(t)}^{b(t)} g(x)\,\mathrm{d}x = g(b)\frac{\mathrm{d}b}{\mathrm{d}t} - g(a)\frac{\mathrm{d}a}{\mathrm{d}t}$$

$$\int u\frac{\mathrm{d}v}{\mathrm{d}x}\,\mathrm{d}x = u\,v - \int v\frac{\mathrm{d}u}{\mathrm{d}x}\,\mathrm{d}x \quad \text{and} \quad \int g(u)\frac{\mathrm{d}u}{\mathrm{d}x}\,\mathrm{d}x = \int g(u)\,\mathrm{d}u$$

$y = \sin\theta \iff \theta = \sin^{-1} y$

$f(x)$	$\dfrac{\mathrm{d}f}{\mathrm{d}x}$	$f(x)$	$\dfrac{\mathrm{d}f}{\mathrm{d}x}$	$f(x)$	$\dfrac{\mathrm{d}f}{\mathrm{d}x}$
x^n	$n\,x^{n-1}$	$\ln x$	$1/x$	$\sinh x$	$\cosh x$
e^x	e^x	$\log_a(x)$	$(x\ln a)^{-1}$	$\cosh x$	$\sinh x$
a^x	$a^x \ln a$	e^{-x^2}	$-2x\,\mathrm{e}^{-x^2}$	$\tanh x$	$\mathrm{sech}^2 x$
$\sin x$	$\cos x$	$\sin^{-1} x$	$(1-x^2)^{-1/2}$	$\sinh^{-1} x$	$(1+x^2)^{-1/2}$
$\cos x$	$-\sin x$	$\cos^{-1} x$	$-(1-x^2)^{-1/2}$	$\cosh^{-1} x$	$(x^2-1)^{-1/2}$
$\tan x$	$\sec^2 x$	$\tan^{-1} x$	$(1+x^2)^{-1}$	$\tanh^{-1} x$	$(1-x^2)^{-1}$

$\mathrm{erf}(-x) = -\mathrm{erf}(x)$

$$\int_0^x \mathrm{e}^{-t^2}\,\mathrm{d}t = \frac{\sqrt{\pi}}{2}\,\mathrm{erf}(x) \quad \text{and} \quad \int_{-\infty}^{\infty} \mathrm{e}^{\mathrm{i}(x-x_0)t}\,\mathrm{d}t = 2\pi\,\delta(x-x_0)$$

$$\mathrm{erf}(\infty) = 1 \qquad \qquad \text{(see Section 2.4.1)}$$

Fourier transforms *(see Sections 2.4 and 2.5.5)*

$$F(k) = \int\limits_{-\infty}^{\infty} f(x)\, e^{-ikx}\, dx \quad \Longleftrightarrow \quad f(x) = \tfrac{1}{2\pi}\int\limits_{-\infty}^{\infty} F(k)\, e^{ikx}\, dk \qquad\qquad \int\limits_{-\infty}^{\infty} |f(x)|\, dx < \infty$$

$f(x)$	$F(k)$	$f(x)$	$F(k)$
$f(x+x_0)$	$e^{ikx_0}\,F(k)$	$\dfrac{df}{dx}$	$ik\,F(k)$
$f(x)\otimes g(x)$	$G(k)\,F(k)$	$f(x)\,g(x)$	$\tfrac{1}{2\pi}\,G(k)\otimes F(k)$
$\delta(x-x_0)$	e^{-ikx_0}	1	$2\pi\,\delta(k)$
$\delta(x+d)+\delta(x-d)$	$2\cos(kd)$	$\delta(x+d)-\delta(x-d)$	$2i\sin(kd)$
$\begin{cases} 1 & \text{if } \|x\|\leqslant w \\ 0 & \text{otherwise} \end{cases}$	$\tfrac{2}{k}\sin(kw)$	$\displaystyle\sum_{n=-\infty}^{\infty}\delta(x-nd)$	$\displaystyle\sum_{m=-\infty}^{\infty}\delta\!\left(k-\tfrac{2\pi m}{d}\right)$
$\begin{cases} e^{-ax} & \text{for } x\geqslant 0 \\ 0 & \text{otherwise} \end{cases}$	$\dfrac{1}{a+ik}$	$\dfrac{a}{\pi(a^2+x^2)}$	$e^{-a\|k\|}$
$\dfrac{1}{\sigma\sqrt{2\pi}}\exp\!\left(-\dfrac{x^2}{2\sigma^2}\right)$	$e^{-\sigma^2 k^2/2}$		

(to the right of the table:) $f(x)=0$ for $x=\pm\infty$; $a>0$

2.7.1 Dimensional analysis

Theoretical analysis involves the use of equations for understanding physical phenomena in a quantitative manner. Since the derivation of the relationships can be mathematically complicated, it's always worth carrying out 'sanity checks' on the formulae before applying them in detailed calculations; this is a good way of detecting algebraic mistakes and typographical errors. A requirement to simplify to familiar or intuitive results in elementary cases is one part of this approach, but the need for 'dimensional consistency' provides an even more basic test.

Physical parameters related to mechanics can be analysed in terms of their associated dimensions of length, L, time, T, and mass, M. Thus velocity, being a displacement per unit time, has dimensions of LT^{-1}; acceleration, being the rate of change of velocity, LT^{-2}; force, from Newton's second law of motion, MLT^{-2}; energy, from $work = force \times distance$, ML^2T^{-2}; and so on. While it may be necessary to add charge, Q, and temperature, Θ, to the basic list for dealing with electromagnetism and thermodynamics, the balance

implied by an = symbol means that the dimensions on both sides
of an equation must match up (or else something has gone wrong).
Indeed, the dimensions of every component separated by a + or −
must be the same, and the arguments of functions, such as exp, log,
sin and cos, should be dimensionless.

Part II

Elastic scattering

The basics of X-ray and neutron scattering

Having covered the preliminaries, we are now in a position to start discussing the basic concepts of X-ray and neutron scattering. For simplicity, we will initially restrict ourselves to the case where there is no exchange of energy in the process.

3.1 An idealized scattering experiment

The scattering of an X-ray photon, or a neutron, by a sample is characterized by the resultant change in its momentum, $\mathbf{P}$, and energy, E. This is shown schematically in Fig. 3.1, where a particle incident with a wavevector $\mathbf{k}_i$ and angular frequency ω_i emerges with a final wavevector $\mathbf{k}_f$ and frequency ω_f. The fundamental relationships of quantum mechanics, eqns (1.12) and (1.13), enable the *momentum transfer* to be expressed as

$$\mathbf{P} = \hbar\,\mathbf{k}_i - \hbar\,\mathbf{k}_f = \hbar\,\mathbf{Q}\,, \qquad (3.1)$$

$$|\mathbf{k}| = \frac{2\pi}{\lambda}$$

where the Planck constant $\hbar = h/2\pi$ and

$$\mathbf{Q} = \mathbf{k}_i - \mathbf{k}_f\,, \qquad (3.2)$$

and, similarly, the *energy transfer* as

$$E = \hbar\,\omega \quad \text{where} \quad \omega = \omega_i - \omega_f\,. \qquad (3.3)$$

$$\omega = 2\pi\nu$$

The momentum and energy gained by the scattered particle is equal to that lost by the sample, of course, and *vice versa*. The definitions of $\mathbf{P}$ and E as 'incident minus final', rather than the other way around, is a matter of convention.

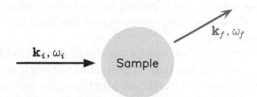

Fig. 3.1 A schematic representation of a particle being scattered by a sample.

$\mathbf{P} = (P_x, P_y, P_z)$

$\mathbf{Q} = (Q_x, Q_y, Q_z)$

An ideal scattering experiment consists of a measurement of the proportion of incident particles that emerge with a given energy and momentum transfer. This is encoded in a four-dimensional function $S(\mathbf{P}, E)$, traditionally called the 'scattering law', where the vector $\mathbf{P}$ has three components; it is frequently written in terms of $\mathbf{Q}$ and ω, as $S(\mathbf{Q}, E)$ and $S(\mathbf{Q}, \omega)$. To keep the discussion as simple as possible in the early stages, we will begin by restricting ourselves to the special case of *elastic* scattering,

$$\boxed{E = 0 \quad \Longleftrightarrow \quad \omega = 0} \,, \tag{3.4}$$

where there is no exchange of energy. This reduces the dimensionality of the problem by one, because the scattering law is then only a function of momentum, or wavevector, transfer:

$$S_{\mathrm{el}}(\mathbf{Q}) = S(\mathbf{Q}, \omega{=}0). \tag{3.5}$$

The condition of eqn (3.4) means that the modulus of the wavevector, and hence the wavelength λ, is unchanged upon scattering,

$$\boxed{|\mathbf{k}_i| = |\mathbf{k}_f| = \frac{2\pi}{\lambda}} \,, \tag{3.6}$$

but the precise reason is slightly different for X-rays and neutrons. For a (massless) photon, the dispersion relation $c = \omega/k$, where c is the speed of light and $k = |\mathbf{k}|$, implies that

$$E_i = \hbar\,\omega_i = \hbar\,c\,|\mathbf{k}_i| \quad \text{and} \quad E_f = \hbar\,\omega_f = \hbar\,c\,|\mathbf{k}_f| \,,$$

which leads to eqn (3.6) following the elasticity requirement that $E_i = E_f$. For a neutron, of mass m_{n}, the energy transfer is given by the change in its kinetic energy

$$E = \frac{|\hbar\mathbf{k}_i|^2}{2m_{\mathrm{n}}} - \frac{|\hbar\mathbf{k}_f|^2}{2m_{\mathrm{n}}} \,,$$

which also yields eqn (3.6) for $E{=}0$.

3.1.1   Elastic scattering and momentum transfer

When talking about a scattering experiment, the picture that comes to mind is of particles in an incident beam being deflected in different directions. How does the geometry of the process relate to the wavevector transfer $\mathbf{Q}$?

The vector diagram for an elastic scattering event is shown in Fig. 3.2, where an incoming particle is deflected through an angle of 2θ; the factor of two is conventional and algebraically convenient. From eqn (3.6), the triangle representing eqn (3.2) is isosceles with the two equal sides being of length $2\pi/\lambda$. Elementary trigonometry then leads to the result

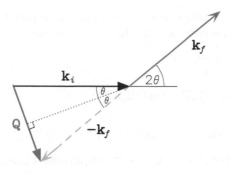

$$\frac{|\mathbf{Q}|}{2} = |\mathbf{k}_i| \sin\theta = \frac{2\pi}{\lambda} \sin\theta$$

Fig. 3.2 The vector diagram for elastic scattering, $|\mathbf{k}_i| = |\mathbf{k}_f|$, through an angle of 2θ.

$$\boxed{Q = \frac{4\pi \sin\theta}{\lambda}} \quad , \qquad (3.7)$$

where $Q = |\mathbf{Q}|$. Although this links the magnitude of the momentum transfer to the wavelength and scattering angle, it doesn't capture all the spatial aspects of the problem.

To define the deflection of the scattered particle completely, a second angle, ϕ, is required for the measurement of rotation about $\mathbf{k}_i$. This is illustrated in Fig. 3.3, where the z-axis has been chosen so that it is parallel to the incident beam:

$$\mathbf{k}_i = \left(0, 0, \frac{2\pi}{\lambda}\right).$$

The angles 2θ and ϕ determine the direction of the outgoing X-ray, or neutron, in much the same way as latitude and longitude give the location of a point on the surface of the Earth (in this case, with respect to the 'north' pole and a reference meridian). More precisely, a careful examination of Fig. 3.3 allows $\mathbf{k}_f$ to be expressed as

$$\mathbf{k}_f = \frac{2\pi}{\lambda} \left(\sin 2\theta \cos\phi, \; \sin 2\theta \sin\phi, \; \cos 2\theta\right),$$

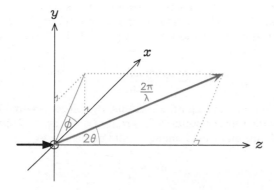

$$0 \leqslant 2\theta \leqslant \pi$$

$$0 \leqslant \phi < 2\pi$$

Fig. 3.3 The scattering geometry in both Cartesian and spherical polar coordinates.

$$\sin 2\theta = 2\sin\theta\cos\theta$$

$$\cos 2\theta = 1 - 2\sin^2\theta$$

where $|\mathbf{k}_f| = 2\pi/\lambda$. Trigonometric double-angle formulae then allow the wavevector transfer, $\mathbf{k}_i - \mathbf{k}_f$, to be written as

$$\mathbf{Q} = \frac{4\pi\sin\theta}{\lambda}\left(-\cos\theta\cos\phi,\ -\cos\theta\sin\phi,\ \sin\theta\right), \qquad (3.8)$$

which satisfies eqn (3.7) as required.

3.1.2 The differential cross-section

The basic quantity that a scattering experiment aims to measure is the fraction of incident particles that emerge in various directions, as defined by the *spherical polar* coordinates 2θ and ϕ in Fig. 3.3. If the neutrons or X-ray photons form a steady stream of incoming particles, then the incident *flux*, Φ, is usually specified by their number per unit time and per unit area perpendicular to the 'flow' (with SI units of $m^{-2}s^{-1}$). Their rate of arrival in the direction of 2θ and ϕ, into a detector that subtends a small *solid angle* of $\Delta\Omega$ *steradians* (sr) at the sample, can similarly be stated as R per unit time (s^{-1}). The ratio

$$\frac{\text{Number deflected by }(2\theta,\phi)\ per\ unit\ solid\ angle}{\text{Number of incident particles }per\ unit\ area\ of\ beam} = \frac{R(2\theta,\phi)/\Delta\Omega}{\Phi}$$

is then closely related to the desired scattering function. It is known as the *differential cross-section*, $\mathrm{d}\sigma/\mathrm{d}\Omega$, and has SI units of $m^2\,sr^{-1}$ or an area per unit solid angle.

The precise definition of the differential cross-section used varies slightly depending on the field of application, but is often quoted per atom, or per molecule, through a division by the number of scattering units of interest, N, in the sample:

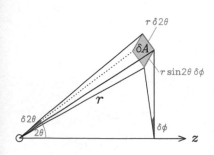

$$\delta\Omega \approx \frac{\delta A}{r^2}$$

$$\approx \sin 2\theta\,\delta\phi\,\delta 2\theta$$

Solid angles

Radians are a dimensionless measure of planar angles, giving the length of the circular arc that is created when a radius is rotated in one dimension. Steradians are the corresponding unit of solid angle, Ω, which give the area of the spherical shell fragment generated when a radius is wiggled in two dimensions:

$$\Omega = \frac{\text{Area of spherical shell fragment}}{\text{Radius}^2}.$$

The solid angle representing all directions, therefore, is equal to the area of a sphere divided by the square of its radius, or 4π sr. An infinitesimally small area $\mathrm{d}A$ subtends an element of solid angle

$$\mathrm{d}\Omega = \frac{\mathrm{d}A}{r^2} \qquad (3.9)$$

at a distance of r, which can be integrated up for finite regions.

$$\boxed{\frac{\mathrm{d}\sigma}{\mathrm{d}\Omega} = \frac{R(2\theta, \phi)}{N\,\Phi\,\Delta\Omega}} \qquad (3.10)$$

where, strictly speaking, the equality holds in the limit of $\Delta\Omega \to 0$. Two common alternative forms of $\mathrm{d}\sigma/\mathrm{d}\Omega$ have a factor of V (m^3) or M (kg) in the denominator of eqn (3.10), instead of N, so that the differential cross-section is normalized by the volume or the mass, respectively, of the sample. This leads to different dimensions and units for $\mathrm{d}\sigma/\mathrm{d}\Omega$, and care must be exercised in ascertaining, or stating, which convention has been employed.

3.1.3 Elastic versus total scattering

The information contained in the differential cross-section can easily be translated into an equivalent $S_{el}(\mathbf{Q})$ of eqn (3.5) as long as all the scattering interactions can be assumed to be elastic and the incident beam contains particles of the same (known) wavelength λ; it simply requires the transformation of the $(2\theta, \phi)$ directions into corresponding values of wavevector transfer with eqn (3.8). While a *monochromatic* beam can usually be arranged, the absence of any energy transfer is the fundamental assumption that is made when equating $\mathrm{d}\sigma/\mathrm{d}\Omega$ with $S_{el}(\mathbf{Q})$.

In reality a sample will exhibit both elastic and *inelastic* ($E \neq 0$) scattering interactions, and they cannot be distinguished without a measurement or a selection of the energy of the outgoing particle. Diffraction experiments generally don't involve any energy discrimination, however, and rely on the property that the scattering tends to be predominantly elastic:

$$\frac{\mathrm{d}\sigma}{\mathrm{d}\Omega} \approx \left(\frac{\mathrm{d}\sigma}{\mathrm{d}\Omega}\right)_{el} = \frac{R_{el}(2\theta, \phi)}{N\,\Phi\,\Delta\Omega}\,. \qquad (3.11) \qquad\qquad R = R_{el} + R_{inel}$$

The small difference between the 'total' and the elastic differential cross-sections, $(\mathrm{d}\sigma/\mathrm{d}\Omega)_{inel}$, is the focus of inelastic studies. For most diffraction work it is just a nuisance that has to be minimized through experimental design, by using low temperatures for example, but is otherwise simply ignored. The exception is liquids and amorphous analysis, where an attempt is made to model and correct for the discrepancy.

3.2 Scattering by a single fixed atom

The above discussion of the differential cross-section was concerned with the description of scattering data. Now let's try to understand how the structure of the sample, at the atomic level, relates to the measured deflections. After all, this theoretical question is at the heart of this book.

The simplest place to start is a single 'fixed' atom. How would it scatter a beam of X-rays or neutrons? Well, a steady stream of particles of wavelength λ travelling in the z direction can be described mathematically by the complex plane wave

$$\psi_i = \psi_{\mathrm{o}}\,e^{ikz}, \tag{3.12}$$

which is invariant with respect to x and y, where the incident flux $\Phi = |\psi_{\mathrm{o}}|^2$, $k = 2\pi/\lambda$ and $i^2 = -1$; the temporal variation of the wave and the phase offset have been subsumed into the argument of ψ_{o}. By a fixed atom we mean that it cannot move or exchange energy in any way. The interaction is elastic, therefore, and the wavenumber of the outgoing particles is also k.

$$\psi_{\mathrm{o}} = |\psi_{\mathrm{o}}|\,e^{i(\phi_{\mathrm{o}} - \omega t)}$$

After interacting with the atom, the particles move radially outwards. If the origin is centred on the atom, so that the wavevector $\mathbf{k}_f$ is parallel to the displacement vector $\mathbf{r}$, then the sinusoidal part of the scattered wave is

$$e^{i\mathbf{k}_f \bullet \mathbf{r}} = e^{ikr},$$

where $r = |\mathbf{r}|$. Unlike the incident beam, which is collimated, the number of outgoing particles per unit area will fall with the distance r according to the inverse-square law. Hence, the scattered wave will take the general form

$$\psi_f = \psi_{\mathrm{o}}\,\mathrm{f}(\lambda, \theta)\,\frac{e^{ikr}}{r}\,, \tag{3.13}$$

for $r > 0$, where the function $\mathrm{f}(\lambda, \theta)$ tells us about the chances that a particle of a given wavelength is deflected in a certain direction (the symmetry of the problem indicating that it will not depend on the angle ϕ); the ψ_{o} ensures that the rate of scattering is proportional to the incident flux. As the characteristics of $\mathrm{f}(\lambda, \theta)$ are different for X-rays and neutrons, this is a useful point to draw the distinctions between them.

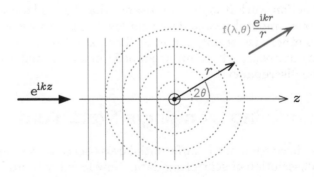

Fig. 3.4 A steady stream of neutrons being scattered elastically and isotropically (with f = constant) by the nucleus of a single fixed atom at the origin.

3.2.1 **Nuclear scattering lengths**

The neutron case is usually straightforward because $f(\lambda,\theta)$ tends to be invariant with respect to wavelength and scattering angle:

$$f(\lambda,\theta) = -b, \qquad (3.14)$$

where the constant b is called a *scattering length* and the minus sign is a matter of convention. The name is apt since $f(\lambda,\theta)$ must have the same dimensions as the r in the denominator of eqn (3.13) for ψ_f to be consistent with ψ_i. The simplicity of eqn (3.14) is subject to certain provisos, the most important of which is that it applies to the scattering of neutrons by the nucleus of the atom; their magnetic interaction with the orbital electrons will be considered shortly in Section 3.2.3.

The isotropic nature of the nuclear scattering stems from the very short range of the related strong interaction by comparison with the wavelength of thermal neutrons: around 10^{-14}m instead of 10^{-10}m. As far as the neutrons are concerned, the scattering is from a 'point source'. The deflections occur equally in all directions, therefore, with the situation being analogous to the diffraction of water waves in a 'ripple tank' by a narrow aperture. The independence from λ implied by eqn (3.14) holds for most nuclei, to a good approximation, in the neighbourhood of thermal wavelengths. The few instances where this is not so, such as for cadmium, gadolinium and boron, are associated with high likelihoods of *absorption* rather than scattering. All nuclei exhibit this phenomenon in the vicinity of certain wavelengths or energies, called *resonances*, but not many do so for the λ of interest; given the rarity, a detailed discussion is delegated to Appendix B for simplicity.

While scattering lengths are complex numbers in principle, their imaginary parts are generally so small that the b can be treated as real. Absorbing nuclei are the exceptions, but eqn (3.14) does not apply to them anyway because both the modulus and argument of $f(\lambda,\theta)$ varies strongly with wavelength close to a resonance. The magnitude of b determines the strength of the scattering, whereas its sign indicates whether the incident and outgoing waves are in or (180°) out of phase.

The scattering interaction quantified by the constant b is not well understood, but it depends on the makeup of the nucleus and the orientation of its spin (if not zero) relative to that of the neutron. This means that the scattering lengths

- are isotope specific,
- have two different values for nuclei with a non-zero spin and
- do not vary with atomic number in a simple or monotonic way.

The nucleus of ordinary hydrogen, for example, which consists of a proton, has a spin of $1/2$ like the neutron. Quantum mechanics shows that two such fermions can be combined to yield a resultant spin of 1 in three different ways and 0 in just one configuration. The scattering lengths for these *triplet* and *singlet* states, denoted by + and −, respectively, are[†]

[†] Koester (1977)

$$b^+ = 1.085 \times 10^{-14}\,\text{m} \quad \text{and} \quad b^- = -4.750 \times 10^{-14}\,\text{m}\,.$$

On average, therefore, a hydrogen nucleus (or a proton) has a scattering length of

$$\langle b \rangle = \tfrac{3}{4}\,b^+ + \tfrac{1}{4}\,b^- = -0.374 \times 10^{-14}\,\text{m} \tag{3.15}$$

and a standard, or *root mean square* (rms), deviation of

$$\Delta b = \sqrt{\left\langle (b - \langle b \rangle)^2 \right\rangle}$$

$$\Delta b = \sqrt{\langle b^2 \rangle - \langle b \rangle^2} = 2.527 \times 10^{-14}\,\text{m} \tag{3.16}$$

where

$$\langle b^2 \rangle = \tfrac{3}{4}\left(b^+\right)^2 + \tfrac{1}{4}\left(b^-\right)^2 = 6.524 \times 10^{-28}\,\text{m}^2\,.$$

Deuterium, which is the heavier isotope of hydrogen with unit spin, by contrast, has scattering lengths of[†]

$$b^+ = 0.953 \times 10^{-14}\,\text{m} \quad \text{and} \quad b^- = 0.098 \times 10^{-14}\,\text{m}$$

The arithmetic of quantum mechanical spin

When a particle of spin S_1 combines with another of spin S_2, the resultant is specified by the *operator* sum

$$\mathbf{S} = \mathbf{S}_1 + \mathbf{S}_2\,.$$

This deceptively simple expression, however, hides a surprising amount of complexity! For our purposes here, it will suffice to say that the total spin S can take integer-spaced values between

$$S_1 + S_2 \geqslant S \geqslant |S_1 - S_2|$$

and that each case is associated with the $2S+1$ possibilities

$$S_z = S,\, S-1,\, S-2,\, \cdots,\, 2-S,\, 1-S,\, -S\,,$$

where the z-axis is traditionally chosen to be the reference direction along which a discriminating magnetic field is applied. Thus, for example, a neutron ($S_1 = 1/2$) can interact with a nucleus of unit spin ($S_2 = 1$) to yield the quartet

$$S = \tfrac{3}{2} \quad \text{with} \quad S_z = \tfrac{3}{2},\, \tfrac{1}{2},\, -\tfrac{1}{2},\, -\tfrac{3}{2}$$

and the doublet

$$S = \tfrac{1}{2} \quad \text{with} \quad S_z = \tfrac{1}{2},\, -\tfrac{1}{2}\,.$$

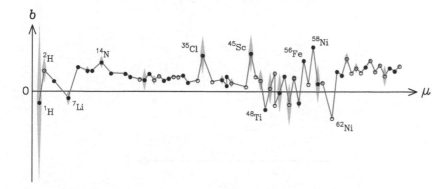

Fig. 3.5 The erratic variation of the neutron scattering lengths, b, with atomic mass, μ, across the periodic table. The average over the nuclear–neutron spin combinations, $\langle b \rangle$, and its standard deviation, Δb, are indicated by the points and shaded regions, respectively; the most abundant isotopes are marked by filled circles, whereas the least common and the absorbing nuclei have been omitted.

where the $+$ and $-$, respectively, refer to the resultant *quartet* and *doublet* states with spins of $3/2$ and $1/2$. On average, therefore, deuterium has

$$\langle b \rangle \;=\; \tfrac{2}{3}\, b^+ + \tfrac{1}{3}\, b^- \;=\; 0.668 \times 10^{-14}\,\mathrm{m} \tag{3.17}$$

and a standard deviation of

$$\Delta b \;=\; \sqrt{\left[\tfrac{2}{3}\,(b^+)^2 + \tfrac{1}{3}\,(b^-)^2 \right] - \langle b \rangle^2} \;=\; 0.403 \times 10^{-14}\,\mathrm{m}\,. \tag{3.18}$$

These examples illustrate both the spin and isotope dependence of neutron scattering lengths; their unpredictable variation through the periodic table can be seen in Fig. 3.5.

3.2.2 Atomic form factors

The characteristics of $f(\lambda, \theta)$ for the scattering of X-rays by an atom are very different to those considered for the neutron, because the relevant interaction is a long-range electromagnetic one with the orbital electrons rather than a short-range force exchanged with its nucleus. Assuming the common situation where the wavelengths do not correspond to energies at which significant absorption occurs, with the resonant case being addressed in Appendix B, $f(\lambda, \theta)$ is a real-valued function which

- diminishes monotonically with increasing θ and decreasing λ,
- has the same sign for all elements and
- has a magnitude proportional to the atomic number Z.

The last two properties are easy to understand since the strength of the scattering depends on the number of orbital electrons, which

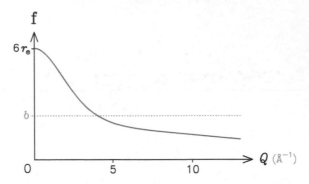

Fig. 3.6 The X-ray form factor for carbon has a maximum proportional to its atomic number, 6, and decays to zero as the wavevector transfer, $Q = 4\pi \sin\theta/\lambda$, increases; r_e is the classical radius of the electron. The equivalent variation of $f(\lambda,\theta)$ for the scattering of neutrons by a ^{12}C nucleus is indicated with a dotted line.

is equal to the number of protons in a neutral atom, and all have the same (negative) charge. The reason for the first feature will become clearer later but, for now, let's go back to the analogy with the diffraction of plane water waves by an aperture in a ripple tank. As the gap in the barrier is widened, the emergent wave pattern changes from semi-circular to more planar with a little curvature at the edges; an increasing proportion of the wave is propagated in the forward direction, therefore, and less at higher angular deflections. This is pretty much what is happening, with the effect becoming more pronounced as the size of the atom becomes larger with respect to the wavelength of the X-rays.

The exact nature of the decay of $f(\lambda,\theta)$ with increasing scattering angle, 2θ, and decreasing wavelength, λ, depends on the element, but it is a function of the ratio $\sin\theta/\lambda$ and has a shape similar to that shown for carbon in Fig. 3.6. In contrast to the invariance of nuclear scattering lengths of eqn (3.14), this fall-off in $f(\lambda,\theta)$ with the modulus of the elastic wavevector transfer, $Q = 4\pi\sin\theta/\lambda$, is known as the atomic *form factor*:

$$f(\lambda,\theta) = Z\,g(Q)\,r_e \qquad (3.19)$$

where Z is the atomic number, $g(Q)$ decays from one at the origin to zero as $Q \to \infty$ and r_e is the classical radius of the electron, or the *Thomson* scattering length,

$$r_e = \frac{e^2}{4\pi\epsilon_o\, m_e c^2} = 2.818 \times 10^{-15}\,\mathrm{m}\,.$$

The parameters for an analytical approximation to $Z\,g(Q)$, consisting of a constant plus the sum of four origin-centred Gaussians of different widths and amplitudes, can be found in Section 6.1.1 of Volume C of the *International Tables for Crystallography* (2006).

Before moving on, we should note that our elemental formulation of scattering in eqns (3.12) and (3.13) is inadequate in one important respect: it does not to take into account the transverse nature of electromagnetic radiation and its associated *polarization*. Like waves on a string, the sinusoidal fluctuations of the electric (and magnetic) field lie in a particular direction in the plane perpendicular to the propagation vector k. This interacts with the orbital electrons and gives rise to an emergent wave whose amplitude, ψ_f, is scaled by an electric *dipole*-dependence on the angles in Fig. 3.3, $P(\theta, \phi)$, as well as the form factor of eqn (3.19). Rather than modifying eqns (3.12) and (3.13), however, it's simpler to compensate for the shortcoming by applying a multiplicative correction, $P^2(\theta, \phi)$, to the intensity, $|\psi_f|^2$, at a later stage:

$$P^2(\theta, \phi) = 1 - \sin^2 2\theta \cos^2 \phi, \qquad (3.20)$$

where the incident electric field is taken as being aligned with the x-axis ($\phi = 0$). For the three most common cases, this 'polarization factor' reduces to

$$P^2(\theta) = \begin{cases} 1 & \text{for } \phi = \pm \pi/2, \\ \cos^2 2\theta & \text{for } \phi = 0 \text{ or } \pi, \\ \frac{1}{2}\left(1 + \cos^2 2\theta\right) & \text{for an unpolarized source.} \end{cases} \qquad (3.21)$$

The formula for the source of unpolarized X-rays, where the incident electric field is oriented randomly for different photons, is obtained by averaging $P^2(\theta, \phi)$ in eqn (3.20) over the angle ϕ (between 0 and 2π radians).

$$\langle P^2(\theta) \rangle = \frac{1}{2\pi} \int\limits_0^{2\pi} P^2(\theta, \phi) \, \mathrm{d}\phi$$

3.2.3   Magnetic form factors

Neutrons also behave like little bar magnets because they possess a magnetic dipole moment, μ_n, by virtue of their quantum mechanical spin:

$$\mu_n = -1.913\,\mu_N, \qquad (3.22)$$

where

$$\mu_N = \frac{e\hbar}{2\,m_p} = 5.051 \times 10^{-27} \text{ J T}^{-1}$$

is called the *nuclear magneton*. This property is anomalous from a classical point of view, where magnetic dipoles are associated with small loops of current, as a neutron has no charge. With a spin of one half, it can adopt only two configurations relative to an external magnetic field: either parallel or antiparallel, known as 'spin up' and 'spin down' respectively. The former state is favoured energetically at very low temperatures, but a random (unpolarized) mix is difficult to avoid under ambient conditions.

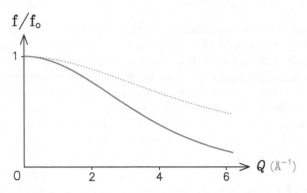

Fig. 3.7 The magnetic form factor for neutrons (in blue) falls off more rapidly with increasing wavevector transfer, $Q = 4\pi \sin\theta/\lambda$, than the X-ray case (dotted grey). The example shown is for the Mn^{2+} ion, with $f_o = f(\lambda, \theta = 0)$.

In addition to the nuclear scattering discussed in Section 3.2.1, a neutron can be deflected through a magnetic interaction between its dipole moment and that of the constituents of the sample. Since the principal source of the latter comes from unpaired electrons in the outer shells of atoms, the magnetic scattering of neutrons

- is ion specific,
- is subject to a form factor and
- does not increase monotonically with atomic number.

The magnetic form factor arises for the same reason as the X-ray case of the preceding subsection, but the fall off with Q is even more pronounced as only the outermost electron orbitals are involved; this can be seen in Fig. 3.7. The parameters for an analytical approximation to the magnetic form factors can be found in Sections 4.4.5 and 6.1.2 of Volume C of the *International Tables for Crystallography* (2006).

When playing with bar magnets as children, we soon discover that the force between them depends on their relative orientations. It is repulsive if neighbouring ends are alike, for example, and attractive when they are of opposite polarity. The same is true at the microscopic level, and results in the magnetic scattering of a neutron by an ion being highly anisotropic. Whether or not this is reflected in the differential cross-section depends on the alignment of the moments of the various atoms of the sample: do they form a correlated pattern or are they oriented randomly? We'll say more about that in Chapter 7, but note here the general property that the effect of a magnetic field is always at right angles to its direction. In terms of a neutron which has been deflected with a wavevector transfer $\mathbf{Q}$, this means that it experienced the component of the magnetization $\mathbf{M}$ that was perpendicular to $\mathbf{Q}$:

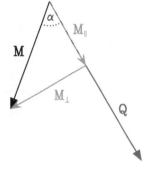

$$\mathbf{M} = \mathbf{M}_{\parallel} + \mathbf{M}_{\perp}$$

$$= \left(|\mathbf{M}| \cos\alpha\right)\widehat{\mathbf{Q}} + \mathbf{M}_{\perp}$$

$$\mathbf{M}_{\perp} = \mathbf{M} - \left(\mathbf{M} \cdot \widehat{\mathbf{Q}}\right)\widehat{\mathbf{Q}}, \qquad (3.23)$$

where $\widehat{Q}$ is a unit vector in the direction of the wavevector transfer. The magnetic interaction can be 'turned off', therefore, if the
moments of all the atoms can be aligned parallel to Q; this may be
achieved through the application of an external magnetic field. An
alternative method of separating the nuclear and magnetic contributions to the scattering is to make measurements above and below
the critical temperature at which magnetic order is destroyed; this
is called the *Curie* or *Néel* temperature, depending on the magnetic
structure of the sample.

$$\widehat{Q} = \frac{Q}{|Q|}$$

Although most neutron experiments are done with an unpolarized
beam, which contains equal numbers of spin up and spin down neutrons, the selective use of a given orientation can be very beneficial
in the study of certain types of magnetic samples. A full polarization analysis is much more challenging, of course, and entails the
additional determination of the spin direction of the outgoing neutron. X-rays also interact with the magnetic elements of a sample,
but their use has been limited until recently owing to the weakness
of the related scattering ($\sim 10^4$ lower rate than that of Section 3.2.2).
The availability of high intensity synchrotron sources through the
1990s has seen a rapid expansion in the field of X-ray magnetic
studies, which provides valuable extra insights; see, for example,
Lovesey and Collins (1996).

3.2.4 Scattering cross-sections

Before moving on to consider scattering from a large assembly of
atoms, let's make the connection between the function $f(\lambda, \theta)$ of eqn
(3.13) and the corresponding *cross-section* σ. The idea of the latter is
quite simple: If the relevant scattering rate from an atom, summed
over all directions, is R (s^{-1} in SI units) when the incident flux is Φ
($s^{-1} m^{-2}$), then this is equivalent to the hit-rate of a classical target
with an area of $\sigma = R/\Phi$ (m^2) perpendicular to the incoming beam.

Rate of scattering = Incident flux × Cross-sectional area .

The number of particles of wavelength λ deflected in the direction of
2θ and ϕ in Fig. 3.3, per unit time and area, is given by the modulus-
squared of eqn (3.13):

$$|\psi_f|^2 = \psi_f\,\psi_f^* = \frac{\Phi}{r^2}\,|f(\lambda, \theta)|^2,$$

where $\Phi = |\psi_o|^2$. Hence the scattering rate, taken over all angles, is
given by the double integral

$$R = \int_{2\theta=0}^{\pi} \int_{\phi=0}^{2\pi} |\psi_f|^2 \, dA,$$

$$dA = r^2 \sin 2\theta \, d\phi \, d2\theta$$

with $\mathrm{d}A = r^2 \sin 2\theta \, \mathrm{d}\phi \, \mathrm{d}2\theta$ being the element of area generated when a radius r pointing in the direction of $(2\theta, \phi)$ is moved through infinitesimally small angular displacements of $\mathrm{d}2\theta$ and $\mathrm{d}\phi$. It leads to the relationship

$$\sigma(\lambda) = 2\pi \int_{2\theta=0}^{\pi} |\mathrm{f}(\lambda, \theta)|^2 \sin 2\theta \, \mathrm{d}2\theta, \qquad (3.24)$$

where the prefactor of 2π comes from the evaluation of the integral with respect to ϕ. For the neutron case of nuclear scattering in eqn (3.14), when $\mathrm{f}(\lambda, \theta)$ is independent of λ and θ, the 2θ-integral readily yields the link between the scattering length b and the corresponding cross-section:

$$\boxed{\sigma = 4\pi |b|^2} \, . \qquad (3.25)$$

1 barn $= 10^{-28} \, \mathrm{m}^2$

A *barn* is a common unit for such areas, and is 10^{28} times smaller than a square metre.

In Section 3.2.1, we noted that the scattering lengths of isotopes whose nuclei have non-zero spin are specified by an average value and a standard deviation,

$$b = \langle b \rangle \pm \Delta b,$$

which is also appropriate for a sample containing a natural mixture of isotopes for a given atom. A rearrangement of this equation, involving squaring and the taking of means, returns the *variance* of eqn (3.16) in the form

$$\langle b^2 \rangle = \langle b \rangle^2 + (\Delta b)^2,$$

and enables the average scattering cross-section to be written as the sum of two quantities:

$$\langle \sigma_{\mathrm{scat}} \rangle = 4\pi \langle b^2 \rangle = \sigma_{\mathrm{coh}} + \sigma_{\mathrm{incoh}}, \qquad (3.26)$$

where the *coherent* and *incoherent* cross-sections are defined by

$$\sigma_{\mathrm{coh}} = 4\pi \langle b \rangle^2 \quad \text{and} \quad \sigma_{\mathrm{incoh}} = 4\pi (\Delta b)^2. \qquad (3.27)$$

(barns)	σ_{coh}	σ_{incoh}
H	1.76	80.27
D	5.59	2.05

We'll discuss the physical significance of σ_{coh} and σ_{incoh} in the next section but, as their names suggest, the former gives rise to interesting structure in the scattering data whereas as the latter generally adds a featureless background signal. The values of $\langle b \rangle$ and Δb calculated in eqns (3.15)–(3.18) show that hydrogen, unlike deuterium, suffers from a very large incoherent cross-section.

In the preceding paragraph, we have implicitly assumed that the scattering lengths are real. While this is usually the case, the imaginary part is not negligible close to an absorption resonance. Whether it be for neutrons or X-rays, the rate of absorption processes can also be characterized by a cross-section, σ_{abs}, which varies strongly with λ in the neighbourhood of a resonance. It is related to the decrease in the intensity, I, of a beam as it passes through a distance l of the corresponding material by

$$I = I_o\, e^{-n\,\sigma_{abs}\,l}, \qquad (3.28)$$

where I_o is the incident intensity and n is the number of atoms per unit volume. The product $n\,\sigma_{abs}$ is called the *absorption coefficient* and its reciprocal is a *mean free path*, or the average distance that a particle travels before being absorbed.

$$n = \frac{\text{Density} \times \text{Avogadro}}{\text{Atomic mass}} = \frac{\rho\, N_A}{m}$$

3.3 Scattering from an assembly of atoms

Having considered the scattering of X-rays and neutrons by an isolated fixed atom, let's see what happens when a large assembly of them is encountered; the latter is, of course, a model for the sample. If the incident beam is represented by a complex plane wave with wavevector $\mathbf{k}_i = (0, 0, k)$, as in eqn (3.12), a particular atom, labelled by the index j, will make a small contribution to the scattered wave, $[\delta\psi_f]_j$, of the form in eqn (3.13):

$$[\delta\psi_f]_j = \psi_o\, e^{i\mathbf{k}_i \bullet \mathbf{R}_j}\, f_j(\lambda, \theta)\, \frac{e^{i\mathbf{k}_f \bullet (\mathbf{r} - \mathbf{R}_j)}}{|\mathbf{r} - \mathbf{R}_j|}, \qquad (3.29)$$

where $f_j(\lambda, \theta)$ defines the relevant interaction characteristics of the jth atom, located at $\mathbf{R}_j$, $\mathbf{k}_f$ is the wavevector of the outgoing particle and $\mathbf{r}$ is an arbitrary position; the geometry is illustrated in Fig. 3.8. Although eqn (3.29) is more complicated than eqn (3.13), because the atom is no longer at the origin, it corresponds to the same physical picture and simplifies to the earlier case when $\mathbf{R}_j = 0$. The

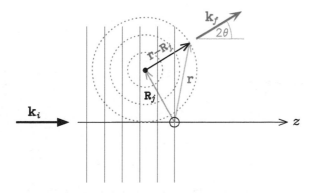

Fig. 3.8 The contribution to the scattered wave from atom j, located at $\mathbf{R}_j$ relative to an arbitrarily defined origin somewhere in the sample.

advantage of this generalization is that it enables us to add up the $\delta\psi_f$ contributions from all N atoms in the sample and work out the net scattered wave ψ_f:

$$\psi_f = \psi_o \, \mathrm{e}^{\mathrm{i}\mathbf{k}_f \bullet \mathbf{r}} \sum_{j=1}^{N} \mathrm{f}_j(\lambda, \theta) \, \frac{\mathrm{e}^{\mathrm{i}\mathbf{Q} \bullet \mathbf{R}_j}}{|\mathbf{r} - \mathbf{R}_j|} \,, \qquad (3.30)$$

where everything not involving j has been taken outside of the summation and $\mathbf{Q} = \mathbf{k}_i - \mathbf{k}_f$ is the wavevector transfer of eqn (3.2). While this is a straightforward application of the superposition principle, there is an implicit assumption that the scattering process is weak. That is to say, the scattered wavelets have a negligible effect on the incident beam. This simplification is often referred to as the *Born*, or the *kinematical*, approximation and holds for both X-rays and neutrons in most instances. Events where particles are deflected more than once have also been ignored, and this too is conditional on very low interaction rates. Depending on the size of the sample, and the geometry of the experimental setup, corrections for multiple scattering and absorption may need to be made, but we will not go into such issues.

The distance to the detectors, where the measurements are taken, is much larger than the typical size of a sample. This means that, to a very good approximation,

$$|\mathbf{r} - \mathbf{R}_j| = |\mathbf{r}| = r \,, \qquad (3.31)$$

with the situation being illustrated in Fig. 3.9. In this Fraunhofer or *far-field* limit, the denominator in eqn (3.30) can be replaced with r and taken outside the summation. The scattered wave then has a modulus-squared of

$$\left| \mathrm{e}^{\mathrm{i}\mathbf{k}_f \bullet \mathbf{r}} \right|^2 = 1$$

$$|\psi_f|^2 = \frac{\Phi}{r^2} \left| \sum_{j=1}^{N} \mathrm{f}_j(\lambda, \theta) \, \mathrm{e}^{\mathrm{i}\mathbf{Q} \bullet \mathbf{R}_j} \right|^2 ,$$

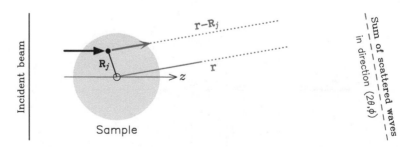

Fig. 3.9 The size of the sample illuminated by the incident beam is usually much smaller than the distance at which scattering measurements are taken, so that $|\mathbf{R}_j| \ll |\mathbf{r}| \approx |\mathbf{r} - \mathbf{R}_j|$ for all j.

where $\Phi = |\psi_{\mathrm{o}}|^2$ is the incident flux. Particles of wavelength λ are deflected elastically into a detector with a small area δA, in the direction of 2θ and ϕ (with respect to Fig. 3.3), at a rate R_{el} per unit time given by

$$R_{\mathrm{el}}(2\theta, \phi) = |\psi_f|^2 \, \delta A = \Phi \, \delta\Omega \left| \sum_{j=1}^{N} \mathrm{f}_j(\lambda, \theta) \, \mathrm{e}^{\mathrm{i}\mathbf{Q} \bullet \mathbf{R}_j} \right|^2 ,$$

where $\delta\Omega = \delta A / r^2$ is the solid angle subtended by the detector at the sample. Hence, the differential cross-section of eqn (3.11) can be related to the structure of the sample, with an atom of type f_j located at $\mathbf{R}_j$, for $j = 1, 2, 3, \ldots, N$, through

$$\boxed{\left(\frac{\mathrm{d}\sigma}{\mathrm{d}\Omega}\right)_{\mathrm{el}} \propto \left| \sum_{j=1}^{N} \mathrm{f}_j(\lambda, \theta) \, \mathrm{e}^{\mathrm{i}\mathbf{Q} \bullet \mathbf{R}_j} \right|^2 ,} \qquad (3.32)$$

$$\lim_{\delta\Omega \to 0} \frac{R_{\mathrm{el}}(2\theta, \phi)}{\Phi \, \delta\Omega}$$

where the proportionality stems from the fact that the right-hand side has not been divided by N, or the mass or volume, or whatever, of the sample. As noted in Section 3.1.2, the choice of normalization varies across disciplines.

3.3.1 Scattering density and Fourier transforms

Much of Section 3.2 was devoted to a discussion of $\mathrm{f}(\lambda, \theta)$, the function used to describe the nature of the scattering of neutrons and X-rays as a result of their various interactions with an atom. The simplest case was that of eqn (3.14), which was attributed to the nucleus being point-like. By the same token, the decay of $\mathrm{f}(\lambda, \theta)$ with increasing $\sin\theta/\lambda$, or Q, in Sections 3.2.2 and 3.2.3 was put down to the orbital electrons occupying a space whose dimension was comparable to the wavelength of the probing X-rays and neutrons. While arguments based on an analogy with the behaviour of water waves in a ripple tank were offered, a deeper understanding was promised later; let us now deliver.

The summary above suggests that the characteristics of $\mathrm{f}(\lambda, \theta)$ are determined by the size and shape of the object responsible for the scattering. The latter can be quantified by a *scattering length density* (SLD) function, $\beta(x, y, z)$, with SI units of m^{-2}, of the appropriate scattering material; nuclear, electron or magnetization, depending on the type of probe and interaction. An infinitesimally small volume $\mathrm{d}V$, at a given location $\mathbf{R}$, will then encapsulate a point source of isotropic scattering with

$$\mathrm{f}(\lambda, \theta) = \beta(\mathbf{R}) \, \mathrm{d}V.$$

The substitution of this elemental form of $\mathrm{f}(\lambda, \theta)$ in the preceding analysis, and a three-dimensional summation over the x, y and z

arguments of $\beta(\mathbf{R})$, leads to the following continuum generalization of eqn (3.32):

$$dV = d^3\mathbf{R} = dx\,dy\,dz$$

$$\left(\frac{d\sigma}{d\Omega}\right)_{el} \propto \left| \iiint_V \beta(\mathbf{R})\,e^{i\mathbf{Q}\bullet\mathbf{R}}\,d^3\mathbf{R} \right|^2 \propto S_{el}(\mathbf{Q})\,, \qquad (3.33)$$

where V is the total volume within which scattering may occur. The elastic differential cross-section is related to the structure of the sample, therefore, through the Fourier transform of its SLD function, $\beta(\mathbf{R})$.

The connection between the SLD distribution of a given atom, j, centred on the origin, $\beta_j(\mathbf{R})$, and the corresponding $f_j(\lambda,\theta)$ becomes apparent when $\beta(\mathbf{R})$ is expressed as the sum of N discrete contributions:

$$\beta(\mathbf{R}) = \sum_{j=1}^{N} \beta_j(\mathbf{R}-\mathbf{R}_j)\,, \qquad (3.34)$$

where $\mathbf{R}_j$ is the location of the jth atom. The substitution of eqn (3.34) into eqn (3.33), and the use of the 'shifting theorem' of Fourier transforms,

$$\iiint \beta_j(\mathbf{R}-\mathbf{R}_j)\,e^{i\mathbf{Q}\bullet\mathbf{R}}\,d^3\mathbf{R} = e^{i\mathbf{Q}\bullet\mathbf{R}_j} \iiint \beta_j(\mathbf{R})\,e^{i\mathbf{Q}\bullet\mathbf{R}}\,d^3\mathbf{R}\,,$$

yields eqn (3.32) with the identification that

$$f_j(\lambda,\theta) = \iiint \beta_j(\mathbf{R})\,e^{i\mathbf{Q}\bullet\mathbf{R}}\,d^3\mathbf{R}\,. \qquad (3.35)$$

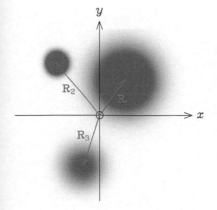

Fig. 3.10 A two-dimensional SLD function, $\beta(x,y)$, consisting of the sum of three discrete components.

Since the result of the integral on the right-hand side is a function of the wavevector transfer $\mathbf{Q}$, which depends on λ, θ and ϕ in general, the absence of ϕ on the left-hand side means that we have assumed an intrinsic symmetry for the SLD distribution of the atoms. Indeed, the properties of Fourier integrals ensure that the simple spherical choice of $\beta_j(\mathbf{R}) = \beta_j(R)$, where $R = |\mathbf{R}|$, leads to $f_j(\mathbf{Q}) = f_j(Q)$ in accordance with the observations noted in Sections 3.2.2 and 3.2.3 that the form factors vary with the ratio of $\sin\theta/\lambda$. The reciprocal relationship between the width of a function and that of its Fourier transform — if one is broad the other is narrow — further explains the relative decays of $f(\lambda,\theta)$ with Q for the various types of scattering interactions discussed in Section 3.2: $\beta(\mathbf{R})$ for a nucleus is essentially a δ-function, and so gives rise to scattering that is uniform with respect to $\mathbf{Q}$; the orbital electrons are spread out, and yield a Fourier transform of finite extent.

3.3.2   Temperature and Debye–Waller factors

So far, we have taken the atoms to be fixed in space. Although this is a reasonable assumption at very low temperatures, with $T \approx 0\,\mathrm{K}$, they oscillate about their equilibrium positions (in a solid) with increasing vigour as the thermal energy rises. The blurred nature of the resultant $\beta(\mathbf{R})$ can be modelled crudely by convolving the zero-temperature SLD function with an isotropic Gaussian whose width is proportional to the square root of the temperature:

$$\beta(\mathbf{R}) \;=\; \Big[\beta(\mathbf{R})\Big]_{T=0} \otimes \gamma \exp\!\left(-\frac{R^2}{\alpha T}\right), \qquad (3.36)$$

$$R^2 = |\mathbf{R}|^2 = x^2 + y^2 + z^2$$

where the constant α, with SI units of $\mathrm{m^2\,K^{-1}}$, defines the variance of the atomic deviations through

$$\langle R^2 \rangle \;=\; \frac{3\alpha T}{2}\,,$$

and $\gamma = (\alpha\,\pi\,T)^{-3/2}$ ensures that the volume integral of $\beta(\mathbf{R})$, or the total amount of scattering material in the sample, does not vary with temperature. In conjunction with eqn (3.33), the convolution theorem of Section 2.4.1 then gives

$$S_{\mathrm{el}}(\mathbf{Q}) \;=\; \Big[S_{\mathrm{el}}(\mathbf{Q})\Big]_{T=0} \times \exp\!\left(-\frac{\alpha T Q^2}{2}\right), \qquad (3.37)$$

and tells us that larger angular deflections are suppressed ever more as the temperature rises. The multiplicative term on the far right, which encodes the exponential fall-off in the scattered intensity with Q^2, is called a *Debye–Waller* factor.

The assumption that all the constituents of a sample vibrate with the same average amplitude, and equally in every direction, was implicit in eqn (3.36). When this is not adequate, because lighter atoms are easier to displace than heavier ones, or the bonding breaks the spatial symmetry, we can consider the effect of temperature on the $f(\lambda, \theta)$ of individual atoms before using eqn (3.32) to obtain $S_{\mathrm{el}}(\mathbf{Q})$. A convolution based argument similar to the above leads to

$$f_j(\lambda, \theta) \;=\; \Big[f_j(\lambda, \theta)\Big]_{T=0} \times \exp\!\left(-\frac{\alpha_j T Q^2}{4}\right), \qquad (3.38)$$

$$Q^2 = Q_x^2 + Q_y^2 + Q_z^2$$

for $j = 1, 2, 3, \ldots, N$. Directionality can be built into the formalism by generalizing the exponents of the Debye–Waller factors to represent ellipsoids in Q_x, Q_y and Q_z instead of spheres, but this requires the introduction of five extra parameters (like α_j) per atom. At high temperatures, the amplitude of the fluctuations can be so large that anharmonicities in the restoring force become apparent and a quadratic dependence on $\mathbf{Q}$ is no longer sufficient. Whatever the case, it's important to note that the decay in the scattered intensity with increasing $\sin\theta/\lambda$ for $T \gg 0$ arises for the same reason as the X-ray atomic and neutron magnetic form factors of Section 3.2: the reciprocal relationship between the width of a function, $\beta_j(\mathbf{R})$, and that of its Fourier transform, $f_j(\mathbf{Q})$.

3.3.3   Coherent and incoherent scattering

In Chapter 2, we saw that Fourier transforms were linear combinations of sinusoidal functions. Whether such a summation gives rise to something interesting depends on the relative weighting with which the different waves are added. In terms of eqn (3.33), $S_{el}(\mathbf{Q})$ only has notable structure if $\beta(\mathbf{R})$ has a certain degree of regularity. For the nuclear scattering of neutrons, where the scattering lengths vary with the isotopic species and its nuclear spin state, it is analytically helpful to decompose the SLD function into an average part which shows a discernible pattern, $\langle\beta(\mathbf{R})\rangle$, and a set of random deviations, $\Delta\beta(\mathbf{R})$:

$$\beta(\mathbf{R}) = \langle\beta(\mathbf{R})\rangle + \Delta\beta(\mathbf{R}) .$$

A one-dimensional example of this procedure is shown in Fig. 3.11. At $T \approx 0\,\mathrm{K}$, the well-defined component of the SLD function can be approximated by

$$\langle\beta(\mathbf{R})\rangle = \sum_{j=1}^{N} \langle\beta_j\rangle\, \delta(\mathbf{R}-\mathbf{R}_j) , \qquad (3.39)$$

where $\langle\beta_j\rangle$ is the spin and, if appropriate, isotope abundance averaged nuclear scattering length of the jth atom, as described in Section 3.2.1, which is located at $\mathbf{R}_j$. The $\Delta\beta(\mathbf{R})$ contribution is different for each neutron, because it depends on the relative orientation of the nuclear and neutron spins, and so

$$\langle\Delta\beta(\mathbf{R})\rangle = 0$$

with the rms fluctuation at $\mathbf{R}_j$ being Δb_j.

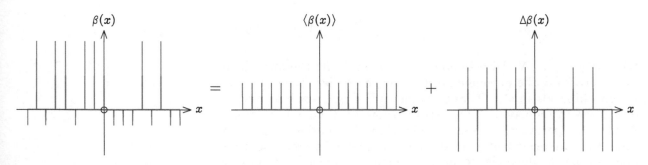

Fig. 3.11 A very simple one-dimensional example of how a function, $\beta(x)$, can be decomposed into the sum of an average pattern showing discernible regularity, $\langle\beta(x)\rangle$, and a set of randomly located fluctuations, $\Delta\beta(x)$.

The Fourier transforms of the one-dimensional example of $\langle \beta(\mathbf{R}) \rangle$ and $\Delta\beta(\mathbf{R})$ in Fig. 3.11, $\langle F(\mathbf{Q}) \rangle$ and $\Delta F(\mathbf{Q})$ respectively, are shown in Fig. 3.12. As anticipated, the former displays notable structure whereas the latter consists of 'white noise'. The resultant (spin averaged) scattering function of eqn (3.33),

$$S_{el}(\mathbf{Q}) \propto \left\langle \left| \langle F(\mathbf{Q}) \rangle + \Delta F(\mathbf{Q}) \right|^2 \right\rangle,$$

can then be expressed as the sum of two terms,

$$S_{el}(\mathbf{Q}) = \left[S_{el}(\mathbf{Q}) \right]_{coh} + \left[S_{el}(\mathbf{Q}) \right]_{incoh}, \tag{3.40}$$

because the cross-term averages to zero:

$$\left\langle \mathcal{R}e \left\{ \langle F(\mathbf{Q}) \rangle^* \Delta F(\mathbf{Q}) \right\} \right\rangle = 0.$$

Explicitly, the structured part of $S_{el}(\mathbf{Q})$, or the coherent scattering, arises from the SLD function of eqn (3.39),

$$\left[S_{el}(\mathbf{Q}) \right]_{coh} \propto \left| \iiint_V \langle \beta(\mathbf{R}) \rangle \, e^{i\mathbf{Q}\cdot\mathbf{R}} \, d^3\mathbf{R} \right|^2, \tag{3.41}$$

whereas the incoherent contribution simply adds a constant background signal with a magnitude equal to the average value of $|\Delta F|^2$; the latter is, in turn, related to the mean incoherent cross-section of the sample (through *Parseval*'s theorem):

$$\left[S_{el}(\mathbf{Q}) \right]_{incoh} \propto \left\langle (\Delta b)^2 \right\rangle. \tag{3.42}$$

Unless otherwise stated, it will be assumed that only the coherent part of the elastic scattering is of interest .

$$|Z_1 + Z_2|^2 = (Z_1 + Z_2)(Z_1 + Z_2)^*$$
$$= |Z_1|^2 + |Z_2|^2$$
$$+ 2\,\mathcal{R}e\,\{Z_1^* Z_2\}$$

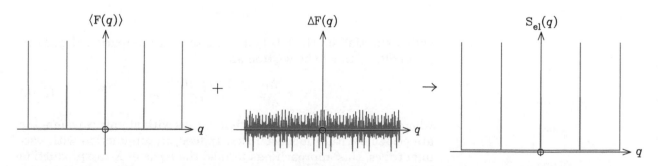

Fig. 3.12 The Fourier transforms of $\langle \beta(x) \rangle$ and $\Delta\beta(x)$ in Fig. 3.11, $\langle F(q) \rangle$ and $\Delta F(q)$ respectively, lead to a resultant scattering function, $S_{el}(q)$, which is proportional to the sum of $|\langle F(q) \rangle|^2$ and a small constant of magnitude $\left\langle |\Delta F(q)|^2 \right\rangle$.

3.3.4   **Mixed scattering events**

Before moving on to consider sources of X-ray and neutron radiation, we should note that the scattering function from a mixture of inter-action events tends to decompose in a manner similar to eqn (3.40). With unpolarized neutrons, for example,

$$S_{el}(\mathbf{Q}) = \big[S_{el}(\mathbf{Q})\big]_{nuc} + \big[S_{el}(\mathbf{Q})\big]_{mag}, \qquad (3.43)$$

where the terms on the right are the independent contributions from nuclear and magnetic scattering.

3.4   **X-rays and synchrotron sources**

In principle, the simplest way of producing X-rays is through the generation of very high temperatures. All objects give out electro-magnetic waves, called *black body* radiation, but the bulk of these emissions occur at a wavelength, λ_{max}, that depends inversely on its temperature T:

$$\lambda_{max} = \frac{2.90 \times 10^{-3}}{T} \text{ m},$$

where T is in kelvin. This is known as *Wein's* Law, and it explains why people (and other warm blooded animals) can be detected with infrared cameras in the dark and why the glow of metals changes from red to white as they are heated. Photons with $\lambda_{max} \approx 1\text{Å}$ would become dominant, therefore, if the temperature could be raised to around 30 million degrees. Since this is comparable to the inside of the sun, and well above the melting point of any material, this is not a very practical avenue.

Rather than trying to convert thermal energy into X-rays, another tack is to make use of kinetic energy. If an electron, having a charge of $-e$, is accelerated towards a metal target by the application of a positive voltage V, then eqn (1.12) and the conservation of energy require that the frequency ν of any emitted photon satisfies

$$h\nu \leqslant eV,$$

with a shortfall on the left being due to heat dissipation. Equation (1.16) allows this to be written as

$$\lambda \geqslant \frac{hc}{eV} \approx \frac{1.24 \times 10^{-6}}{V} \text{ m}, \qquad (3.44)$$

where V is in volts, and shows that X-rays with atomic wavelengths are possible with a modest $10\,\text{kV}$. Indeed, in conjunction with vac-uum tubes, this approach has formed the basis of X-ray production in laboratories for over a century.

The actual conversion of the electron's kinetic energy into a resul-tant photon on striking the metal target arises from two different

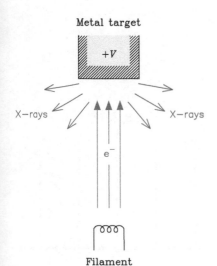

mechanisms. The first is direct, and occurs when an electron (or any charged particle) decelerates rapidly. This is called *Bremsstrahlung*, or 'braking radiation', and yields a continuum of wavelengths with a lower bound given by the equality in eqn (3.44). The second process is indirect, and is initiated by the incoming particle knocking out an electron from one of the inner shells of an atom in the target. This is followed by the relaxation of an electron from a higher energy level, E_j, to the more stable vacant position, E_i, and is accompanied by the emission of a photon with a frequency, ν_{ij}, corresponding to the change in energy:

$$h\nu_{ij} = E_j - E_i .$$

The quantized nature of the atomic orbitals means that this mechanism produces photons at a discrete set of wavelengths,

$$\lambda_{ij} = \frac{hc}{E_j - E_i} , \tag{3.45}$$

subject to the constraint of eqn (3.44). The spectrum of X-rays that emerge is illustrated schematically in Fig. 3.13. By far the strongest feature is the K_α line, at λ_{12}, which results from an electronic transition from the second lowest energy level to the ground state. Hence, this is the primary wavelength at which everyday X-ray work is done in laboratories.

Although vacuum tube technology has been the mainstay of X-ray studies for a century, certain intrinsic features have kept improvements modest. For example, since most of the input kinetic energy is converted into heat rather than photons, the rate at which the target can be cooled restricts the output intensity of the source. The divergence of the X-ray beam and its fixed wavelength (for a given metal target) also impose experimental limitations. A huge step forward has come with the development of dedicated X-ray *synchrotron* facilities.

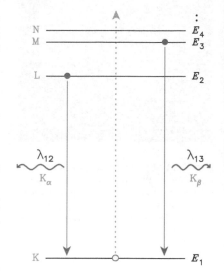

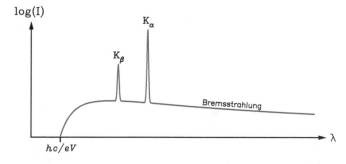

Fig. 3.13 A schematic illustration of the the logarithm of the intensity versus wavelength for an X-ray tube. The sharp lines in the K-series involve electronic transitions down to the lowest energy level, with a principal quantum number of $n = 1$; the subscripts α, β and so on denote changes of $\Delta n = 1, 2, \ldots$, respectively.

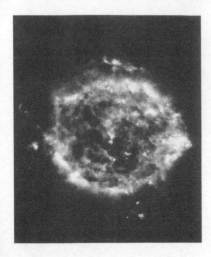

Fig. 3.14 Synchrotron emission from *Cassiopeia A*, the supernova remnant of a massive star that exploded about 300 years ago. (Courtesy of Prof. S. F. Gull, Mullard Radio Astronomy Observatory, Cambridge.)

When a charged particle is accelerated, it emits electromagnetic waves. Bremsstrahlung is an example that involves deceleration. Since acceleration can refer to a change in the direction of motion as much as speed, electrons moving in circular (or helical) orbits in response to a magnetic field radiate; at relativistic speeds, the emissions are in the X-ray waveband. This is the basis of synchrotron radiation. It occurs naturally in space, being responsible for the radio emission from the Galaxy, supernova remnants and extragalactic sources, but was first noticed in circular particle accelerators built for high energy physics in the late 1940s; this is the origin of the pronoun 'synchrotron' for such radiation.

The early work on synchrotron X-rays was done parasitically at facilities designed to study subatomic structure. In this context, the radiation is an undesirable byproduct whereby some of the electrical energy used to speed up the electrons and positrons (or protons and anti-protons) is lost in the process of confining their trajectories to a closed orbit with a magnetic field. As this leakage is minimized by reducing the curvature of the path, storage rings for high energy physics experiments have become ever larger; the LHC, in Geneva, has a circumference of 27 km! Conversely, a higher radius of curvature is advantageous if the principal objective is the production of X-ray synchrotron radiation. Modern sources typically have a diameter of a few hundred metres, with the size being dictated by the number of beamlines planned for the facility.

The great success of synchrotron X-rays is primarily down to their highly collimated nature, which allows the rate at which photons impinge on a sample to be several orders of magnitude higher than that from vacuum tubes. This enormous enhancement is due largely to *relativistic beaming*, whereby photons are radiated preferentially in the (instantaneous) forward direction as the electrons circulate in the storage ring at very close to the speed of light c. A classical analogy of this phenomenon is given by a sprinkler that squirts water uniformly as illustrated in Fig. 3.15(a). From the perspective of someone moving towards the fountain, however, the outflow is elongated in the direction of the relative motion; this is shown in Fig. 3.15(b). Although this familiar *Galilean* transformation breaks down at relativistic speeds, when $v \to c$ and the *Lorentz* factor

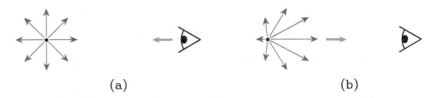

(a)   (b)

Fig. 3.15 A simple classical analogue of relativistic beaming: (a) the uniform outflow of water with respect to the fountain; (b) the corresponding velocity distribution relative to an approaching observer, distorted in the direction of the relative motion.

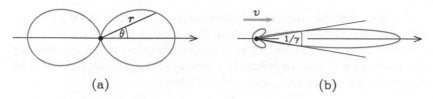

(a)   (b)

Fig. 3.16 Polar diagrams of the intensity, r, versus angle, θ, of the radiation emitted by a charge accelerated tangentially to its direction of motion from the perspective of (a) the rest frame of the charge and (b) an observer in the laboratory.

$$\gamma = \left(1 - \frac{v^2}{c^2}\right)^{-1/2} \gg 1, \qquad (3.46)$$

a Lorentz transformation of the dipole radiation emitted by an accelerated charge in its own rest frame explains why an observer in the laboratory sees a narrow beam of photons aligned with the forward direction of the particle. The case of $\gamma = 3$, corresponding to 94.3% of the speed of light, is illustrated in Fig. 3.16. The bulk of the emission is confined to an angular divergence of $1/\gamma$ radians, typically less than one minute of arc with $\gamma \approx 10^4$, which sweeps tangentially around the circumference of the storage ring like a highly focused search light.

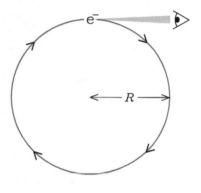

The main characteristics of the radiation emerging from a circular synchrotron are determined by the kinetic energy of the electrons, E, and radius of the orbit, R. The Lorentz factor, for example, which controls the collimation of the beam, is given by

$$\gamma = 1 + \frac{E}{m_e c^2} \approx 1960 E, \qquad (3.47)$$

where the numerical approximation on the far right assumes that $E \gg m_e c^2$ and is given in the common, but non SI, units of GeV. A heavier charged particle, such as a proton, would require the input of much more energy (about a thousand fold) to achieve the same value of γ as an electron. An observer in the plane of the orbit sees brief flashes of duration

$$\Delta T \approx \frac{1}{2\gamma^2} \times \left(\frac{1/\gamma}{2\pi}\right) T \approx \frac{R}{2c\gamma^3},$$

where the period $T \approx 2\pi R/c$ and the $2\gamma^2$ term is the *Doppler* factor for a source of radiation approaching at relativistic speed. This gives rise to a continuous spectrum of photons with a lower bound of $\lambda_{\min} \approx c\,\Delta T$ and a 'critical', or *median*, wavelength of

$$\lambda_c = \frac{4\pi R}{3\gamma^3} \approx \frac{0.56 R}{E^3}\ \text{Å}, \qquad (3.48)$$

where the numerical approximation assumes that the radius is given in metres and the energy in GeV.

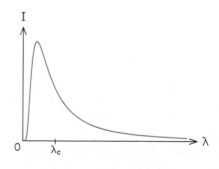

The final notable feature of synchrotron radiation is its intrinsic polarization. Indeed, the detection of a high degree of polarization from an astronomical source is a good indication of an underlying synchrotron process. In terms of a storage ring, the electric field of the emitted photons lies in the orbital plane.

3.4.1   Wigglers and undulators

The picture of a perfectly circular synchrotron considered above is an idealized one. In practice, a storage ring consists of a combination of curved regions, where a uniform magnetic field is applied perpendicular to the orbital plane, and straight sections, containing *magnetic quadrupoles* that focus the electron beam and *RF* (radio frequency) *cavities* which pulse a voltage along their path. The latter give an electrostatic speed boost to the electrons, to compensate for their gradual slowing down as a result of the energy lost in the radiative process. The *bending magnets* are solely responsible for the synchrotron emission in this setup, and our discussion has so far been based on them.

In modern X-ray facilities, known as third generation sources, to distinguish them from the dedicated but conventional second generation ones, even greater intensity and versatility is achieved through the use of insertion devices called *wigglers* and *undulators*. These consist of banks of magnets, of alternating vertical polarity, that are placed in the straight sections of the storage ring to force the electrons to execute horizontal oscillations. The curvature of the zig-zag path induces the synchrotron beam, but its overall fixed orientation allows the emissions to build up in that direction rather than just flash by as with a bending magnet. The perturbation of the linear trajectory is much smaller for undulators than wigglers, and this results in a more focused beam with the power concentrated around certain well-defined, and tunable, wavelengths; details can be found in Als-Nielsen and McMorrow (2001). For these reasons, undulators have become the principal source of synchrotron radiation at X-ray facilities.

The huge improvement in the power of X-ray sources can be seen from Fig. 3.17, where the average brightness from vacuum tubes, bending magnets and undulators is marked on a logarithmic scale; for comparison, the values for three everyday light sources are also indicated. The formal measure of source quality is called *brilliance*, and is defined by

$$\text{Brilliance} = \frac{\text{Number of photons}/\text{second}}{(\text{mrad})^2 \, (\text{mm}^2 \text{ source area}) \, (0.1\% \text{ bandwidth})}$$

where the denominator takes into account the cross-sectional area from which the emissions occur, the solid angular divergence of the beam and the spectral distribution of the photons. Although undulators have opened up avenues of experimental X-ray science that

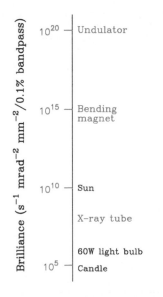

Fig. 3.17 Average brightness values.

were inconceivable until the late 20th century, the enormous level of flux is not without its challenges: the primary optical components, such as *monochromators*, need to be cooled cryogenically and organic samples suffer radiation damage rapidly.

3.5 Reactors and pulsed neutron sources

Outside a nucleus, a neutron decays with a half-life of about 15 minutes into a proton and an electron (and an anti-neutrino):

$$n \rightarrow p + e^- + \bar{\nu}_e .$$

While this process can occur within an atom, leading to the ejection of the electron as β-radiation and the transmutation of the element, the rate is usually much slower. This stability means that atomic nuclei are ultimately the source of neutrons. The question is how to get them out?

Neutrons were discovered by Chadwick, in 1932, following the irradiation of beryllium with α-particles from polonium (^{210}Po). The main nuclear reaction involved is

$$\alpha + {}^9\text{Be} \rightarrow {}^{12}\text{C} + n ,$$

where α-particles are just ^{4}He nuclei. The emission rates from such (α, n) processes are typically 10^6 to 10^8 neutrons per second, which is too low for most experimental work. A better source of neutrons is provided by *fission*, whereby a heavy nucleus splits into two lighter ones and generates some residual γ-rays, neutrons and other subatomic particles. Uranium enriched with its less common isotope, ^{235}U, is the principal raw material. After the initial absorption of a stray neutron, created by an occasional spontaneous fission or the interaction between a cosmic ray and matter, the resultant ^{236}U is unstable and breaks up in a number of different ways. One possible mechanism is

$$n + {}^{235}\text{U} \rightarrow {}^{236}\text{U}^* \rightarrow {}^{134}\text{Xe} + {}^{100}\text{Sr} + 2n ,$$

but the alternatives combine to release an average of 2.5 neutrons per event. The net surplus forms the basis of a self-sustaining chain reaction, where some of the ejected neutrons induce the fission of other ^{235}U atoms, that allows research *reactors* to yield a neutron flux of up to $10^{15}\,\text{s}^{-1}\text{cm}^{-2}$.

Uranium undergoes fission with bombardment by slow neutrons, but those liberated in the nuclear reaction are too fast by a factor of over a thousand. They are made to lose kinetic energy through collisions with light atoms, since the exchange is most efficient between particles of similar mass, to create the right conditions for a chain reaction. The resulting distribution of neutron speeds approaches the Maxwell–Boltzmann function of Fig. 1.13 corresponding to the temperature of the moderating material; the latter, usually graphite or heavy water (D_2O), is kept cool at $T \approx 300\,\mathrm{K}$ with an appropriate heat transfer system. Luckily, the thermal energy neutrons that are best suited to encourage fission are also optimal for use in the study of condensed matter at the atomic and molecular level. If either longer or shorter wavelengths are required preferentially, they can be obtained through further moderation outside the reactor core at the desired temperature; not surprisingly, the outputs are known as hot and cold sources.

Elastic collisions and moderation

The moderation process, where a fast neutron is slowed down by an elastic collision with a slower nucleus, can be understood through a simple example from billiard ball dynamics: a particle of mass m and speed u strikes one that is stationary and of mass M head on; after the collision, their velocities are v and V, respectively, as shown.

Before   After

The relationship between u and v can be ascertained in terms of m and M by applying the principles of the conservation of momentum and kinetic energy. The former requires that

$$mu = mv + MV, \tag{3.49}$$

whereas (twice) the latter implies

$$mu^2 = mv^2 + MV^2. \tag{3.50}$$

After substituting for V from eqn (3.49) in (3.50), the resultant quadratic equation for the ratio v/u can be solved to give

$$\frac{v}{u} = \frac{m - M}{m + M}. \tag{3.51}$$

The alternative result of $u = v$, and $V = 0$, is not physically meaningful as m would then need to pass through M! The ratio of v/u is negative if $m < M$, meaning that m bounces backwards, while both particles move forward when $m > M$. The speed of the incident particle is smallest after the encounter, in fact zero, when $m = M$. Hence eqn (3.51) supports the assertion that the energy exchange in collisions is greatest for particles of similar mass, and explains why neutron moderators consist of light, if not hydrogenous, materials.

The scenario above is called an elastic collision because kinetic energy is conserved in the impact. From the perspective of mass m, however, it is an inelastic scattering event since $|u| \neq |v|$.

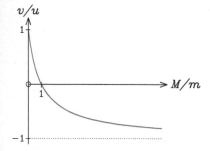

An alternative source of neutrons, which has become increasingly important since the 1990s, is based on accelerator technology similar to that found at synchrotron facilities. In essence, a stream of high energy protons (~ 1 GeV) strikes a target made from a heavy element, such as tungsten or mercury, which results in neutrons and protons (and other subatomic byproducts) being knocked out of the nuclei in a process called *spallation*. Although this debris of ejected nucleons is too hot to be moderated effectively, less energetic neutrons 'evaporate' from the remnant nuclei when they relax from their excited states; these secondary emissions are more amenable to thermodynamic taming.

Spallation sources have a big political advantage over nuclear reactors, in that their construction is far less controversial. This is mainly because they don't need fissile fuel, as it's not a requirement of the target material. There is also no danger of a runaway chain reaction, or meltdown, as the spallation process stops when power is cut off to the accelerator.

3.5.1 The time-of-flight technique

Unlike conventional nuclear reactors, which have a continuous output, neutrons at spallation sources are produced in short bursts as focused bunches of protons are made to hit the target periodically. This is illustrated schematically in Fig. 3.18, where the contribution from the under-moderated, or *epithermal*, neutrons has been omitted for clarity; this fast component would appear at very short times after each pulse, at integer values of τ. The repetition rate, $1/\tau$, depends on the facility, but is typically between 10 and 60 Hz. The lower value is better suited to the longer wavelength neutrons emerging from a cold moderator (20 K rather than 300 K, say), as this allows sufficient time between the pulses for these slower particles to traverse the experimental apparatus.

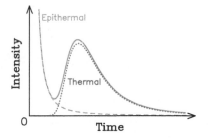

Scattering measurements at reactors and synchrotrons are generally carried out with a beam containing particles of a single (known)

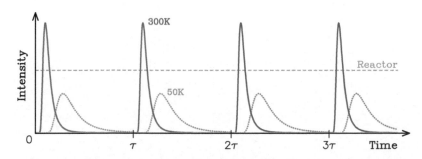

Fig. 3.18 A schematic illustration of the flux of thermal neutrons, from moderators at 300 K and 50 K, at a spallation source with a repetition rate of $1/\tau$, as compared with the uniform temporal output of a conventional nuclear reactor.

wavelength. As noted in Section 3.13, this enables the angular deflections $(2\theta, \phi)$ to be transformed into the corresponding wavevector transfer **Q**, with eqn (3.8), for elastic interactions. Such a monochromation is inefficient, however, since most of the incident flux is lost in the selection procedure. It is unnecessary at a spallation source, because the common starting point of all the neutrons, following the proton pulse, means that the wavelength of each can be determined from basic kinematics and eqn (1.13):

$$\lambda = \frac{h\,(t+t_{\mathrm{o}})}{m_{\mathrm{n}}\,(L+L_{\mathrm{o}})} \,, \tag{3.52}$$

where t is the time, measured from the target impact, at which an elastically scattered neutron is detected at a given flight path distance L from the moderator; t_{o} and L_{o} are small offsets that have to be calibrated. This is the basis of the *time-of-flight* technique. The number of deflections can be normalized against the spectral distribution of the incident neutrons, obtained in the same way by placing monitors in the multichromatic 'white' beam prior to the sample.

Surfaces, interfaces and reflectivity

For our first example of elastic scattering, let's consider the case of *specular reflectivity*. This type of experiment is used in the study of layered materials, and is a good starting point because it simplifies to a one-dimensional problem.

4.1 Reflectivity and Fourier transforms

The geometrical setup for specular reflectivity is shown in Fig. 4.1: a beam of X-rays, or neutrons, of wavelength λ impinges at a grazing angle θ on a planar sample, and the fraction that bounce back with the same attributes is ascertained. These measurements are repeated for different angles at a given wavelength, or the other way around with a multichromatic source, and yield the reflectivity curve $R(Q)$, where

$$Q = \frac{4\pi \sin \theta}{\lambda} \qquad (4.1)$$

and, by definition, $0 \leqslant R \leqslant 1$. The central question is: How does the structural composition of the sample, as encapsulated by its three-dimensional SLD function $\beta(\mathbf{r})$, relate to $R(Q)$?

We should begin by emphasizing that our interest here is purely in layered samples. That is to say, to a good approximation, their

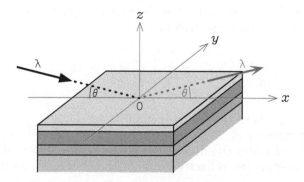

Fig. 4.1 A schematic illustration of an X-ray or neutron, of wavelength λ, being reflected specularly, at an angle θ, from the surface of a layered sample.

SLD is invariant in the x–y plane:

$$\beta(\mathbf{r}) = \begin{cases} \beta(z) & \text{for } |x| < L_x \text{ and } |y| < L_y, \\ 0 & \text{otherwise}, \end{cases} \tag{4.2}$$

where the incident illumination, of the X-rays or neutrons, has been taken to be over a finite rectangular region, $|x| < L_x$ and $|y| < L_y$. In the limit of weak scattering and the Fraunhofer regime, as discussed in Section 3.3, the elastic differential cross-section is given by the Fourier transform of the SLD function:

$$\left(\frac{d\sigma}{d\Omega}\right)_{\text{el}} \propto \left| \iiint_V \beta(\mathbf{r}) \, e^{i\mathbf{Q}\bullet\mathbf{r}} \, d^3\mathbf{r} \right|^2,$$

$$d^3\mathbf{r} = dx \, dy \, dz$$
$$\mathbf{Q}\bullet\mathbf{r} = x \, Q_x + y \, Q_y + z \, Q_z$$

where V is the total volume within which scattering can occur. The substitution of $\beta(\mathbf{r})$ from eqn (4.2), and the evaluation of the separable integrals with respect to x and y along the lines of Section 2.5.2, leads to

$$\left(\frac{d\sigma}{d\Omega}\right)_{\text{el}} \propto 16 \, \frac{\sin^2(L_x Q_x)}{Q_x^2} \, \frac{\sin^2(L_y Q_y)}{Q_y^2} \left| \int_{-\infty}^{\infty} \beta(z) \, e^{i z Q_z} \, dz \right|^2. \tag{4.3}$$

The Q_x and Q_y prefactor on the right-hand side is a product of sinc-squared functions, identical to that for k_x and k_y in Fig. 2.19(a) with $D_x = 2 L_x$ and $D_y = 2 L_y$. It has a maximum when

$$Q_x = 0 \quad \text{and} \quad Q_y = 0, \tag{4.4}$$

of value $16 \, L_x^2 \, L_y^2$, and is concentrated in the narrow region

$$|Q_x| < \frac{\pi}{L_x} \quad \text{and} \quad |Q_y| < \frac{\pi}{L_y}. \tag{4.5}$$

For a sample consisting of planar layers, therefore, most of the incident beam is scattered close to the specular condition of eqn (4.4). In this limit, eqn (4.3) reduces to

$$Q^2 = Q_x^2 + Q_y^2 + Q_z^2$$

$$\left(\frac{d\sigma}{d\Omega}\right)_{\text{el}} \propto 16 \, L_x^2 \, L_y^2 \left| \int_{-\infty}^{\infty} \beta(z) \, e^{-i z Q} \, dz \right|^2, \tag{4.6}$$

with $Q_z = -Q$ given the geometry of Fig. 4.1.

The dependence of $R(Q)$ on $\beta(z)$ is closely related to eqn (4.6). To see this, consider the definitions of the elastic differential cross-section and specular reflectivity. The former was given in Section 3.1.2,

$$\frac{d\sigma}{d\Omega} = \frac{\text{Number deflected by } (2\theta, \phi) \text{ \textit{per unit solid angle}}}{\text{Number of incident particles \textit{per unit area of beam}}},$$

with the interactions being implicitly taken as elastic, and the latter
is simply

$$R = \frac{\text{Rate of specular reflective scattering}}{\text{Rate of incidence}}.$$

A comparison shows that the denominators are linked through the
area of the sample illuminated perpendicular to the incoming beam,
$4 L_x L_y \sin\theta$, whereas the numerator of $d\sigma/d\Omega$ has to be integrated
over a small solid angle, $\Delta\Omega$, in the neighbourhood of the specular
condition to be equated with R:

$$R(Q) = \frac{1}{4 L_x L_y \sin\theta} \iint\limits_{\Delta\Omega} \left(\frac{d\sigma}{d\Omega}\right)_{el} d\Omega. \tag{4.7}$$

This double integral can be evaluated by exploiting the compact nature of the differential cross-section of eqn (4.3), by multiplying its
limiting form in eqn (4.6) with the solid angle corresponding to the
central portion of the region enclosed by eqn (4.5):

$$\iint\limits_{\Delta\Omega} \left(\frac{d\sigma}{d\Omega}\right)_{el} d\Omega \approx \frac{\Delta\Omega}{4} \times \left(\frac{d\sigma}{d\Omega}\right)_{el}\Bigg|_{Q_x=0, Q_y=0, Q_z=-Q}$$

where, after a consideration of the kinematics of elastic scattering
in the vicinity of specular reflection, it can be shown that

$$\Delta\Omega \approx \frac{16\,\pi^2 \sin\theta}{L_x L_y Q^2}. \tag{4.8}$$

This approximation to the integral in eqn (4.7) leads to

$$R(Q) \propto \frac{16\,\pi^2}{Q^2} \left| \int_{-\infty}^{\infty} \beta(z)\, e^{-izQ}\, dz \right|^2, \tag{4.9}$$

where the exact area of the illumination conveniently cancels out
but has been assumed to be large compared to the wavelength of the
radiation.

The proportionality of eqn (4.9) should, in fact, be an equality. The
former was introduced to compensate for the variety of conventions
in use for the normalization of the differential cross-section, but it's
unnecessary with the scattering length density. The SLD is easiest
to calculate for a uniform slab of a single element, especially for the
case of neutron scattering by the nucleus. It's just the product of the
coherent scattering length, $\langle b \rangle$, and the number of atoms per unit
volume, n:

$$\beta = n \langle b \rangle. \tag{4.10}$$

$$\mathbf{Q} = \mathbf{k}_i - \mathbf{k}_f = (-\epsilon, -\delta, 2k_z - \xi)$$

$$|\mathbf{k}_i| = \sqrt{k_x^2 + k_z^2} = \frac{2\pi}{\lambda}$$

$$\left|\widehat{\mathbf{k}}_f\right|^2 = 1$$

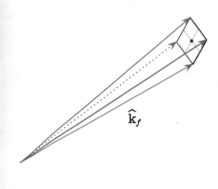

$$\widehat{\mathbf{k}}_f$$

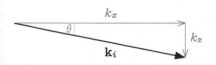

The angular spread of specular reflections

Although specular reflection is defined by $Q_x = 0$, $Q_y = 0$ and $Q_z = -Q$, with reference to Fig. 4.1, the finite size of the illuminated sample means that there is appreciable scattering in the neighbourhood of this condition. Over what solid angle, $\Delta\Omega$, does it spread?

Taking the incident beam to be parallel to the x-axis, the incoming and outgoing wavevectors, $\mathbf{k}_i$ and $\mathbf{k}_f$ respectively, are given by

$$\mathbf{k}_i = (k_x, 0, k_z) \quad \text{and} \quad \mathbf{k}_f = (k_x + \epsilon, \delta, \xi - k_z),$$

where non-zero values of ϵ, δ and ξ characterize departure from the specular condition. The requirement of elastic scattering, $|\mathbf{k}_i|^2 = |\mathbf{k}_f|^2$, imposes the restriction

$$\epsilon^2 + 2\epsilon k_x + \delta^2 + \xi^2 - 2\xi k_z = 0.$$

For small deviations, when the quadratic terms ϵ^2, δ^2 and ξ^2 are negligible, this implies $\xi \approx \epsilon k_x / k_z$. Hence, the unit vector along the scattered beam, $\widehat{\mathbf{k}}_f$, is a function of the independent variables ϵ and δ:

$$\widehat{\mathbf{k}}_f(\epsilon, \delta) \approx \frac{\left(k_x + \epsilon, \delta, \left[\frac{k_x}{k_z}\right]\epsilon - k_z\right)}{\sqrt{k_x^2 + k_z^2}}.$$

$\Delta\Omega$ is then the area of the spherical shell generated by the vertex of $\widehat{\mathbf{k}}_f$ as ϵ and δ take on their allowed range of values in the vicinity of $\epsilon = 0$ and $\delta = 0$.

The extent of the local departure from specular reflection is given by the boundaries of the rectangular region in eqn (4.5). These limits on Q_x and Q_y are equivalent to $\epsilon = \pm\pi/L_x$ and $\delta = \pm\pi/L_y$, respectively, and lead to an estimate of $\Delta\Omega$ through the area of the (small) parallelogram generated by the four corresponding $\widehat{\mathbf{k}}_f(\epsilon, \delta)$ corners. Making use of the vector properties outlined in Section 2.1.1, and that of the cross product in particular,

$$\Delta\Omega \approx \left| \left[\widehat{\mathbf{k}}_f\left(\tfrac{\pi}{L_x}, 0\right) - \widehat{\mathbf{k}}_f\left(\tfrac{-\pi}{L_x}, 0\right)\right] \times \left[\widehat{\mathbf{k}}_f\left(0, \tfrac{\pi}{L_y}\right) - \widehat{\mathbf{k}}_f\left(0, \tfrac{-\pi}{L_y}\right)\right] \right|$$

$$\approx \frac{1}{k_x^2 + k_z^2} \left| \left(\tfrac{2\pi}{L_x}, 0, \tfrac{k_x}{k_z}\tfrac{2\pi}{L_x}\right) \times \left(0, \tfrac{2\pi}{L_y}, 0\right) \right|$$

$$\approx \frac{4\pi^2}{L_x L_y (k_x^2 + k_z^2)} \left| \left(-\tfrac{k_x}{k_z}, 0, 1\right) \right|.$$

The evaluation of the modulus on the far right-hand side, and some algebraic rearrangement, allows the solid angle to be written as

$$\Delta\Omega \approx \frac{4\pi^2 \sin\theta}{L_x L_y k_z^2},$$

where $\sin\theta = k_z/\sqrt{k_x^2 + k_z^2}$ from elementary trigonometry. Finally, the expression for $\Delta\Omega$ given in eqn (4.8) follows from the specular wavevector transfer $Q_z = 2k_z = -Q$.

Silicon, for example, which is often used as a *substrate* upon which layers of sample material are deposited, has $\langle b \rangle = 4.15 \,\text{fm}$, an atomic mass $m = 28.1 \,\text{g mol}^{-1}$ and a density $\rho = 2.33 \,\text{g cm}^{-3}$; this yields an SLD of

$$\beta = \frac{\rho N_A}{m} \langle b \rangle = 2.1 \times 10^{-6} \,\text{\AA}^{-2},$$

where the Avogadro constant $N_A = 6.02 \times 10^{-23} \,\text{mol}^{-1}$ and the values of $\langle b \rangle$ and ρ were multiplied by 10^{-5} and 10^{-24}, respectively, to have consistent length units of Å.

Rather than consisting of a single element, most layers are molecular. If the chemical formula indicates that there are K types of atoms, of which $\#_j$ are of variety j with scattering length $\langle b \rangle_j$, then the SLD is given by

$$\boxed{\beta = \frac{\rho N_A}{m} \sum_{j=1}^{K} \#_j \langle b \rangle_j} \,, \tag{4.11}$$

where $m = \sum \#_j m_j$ is the mass of the molecule and ρ is the bulk density. For the oxide layer which builds up on exposed surfaces of silicon, SiO_2,

$$\beta = \frac{\rho N_A}{m} \left(\langle b \rangle_{Si} + 2 \langle b \rangle_O \right) = 3.5 \times 10^{-6} \,\text{\AA}^{-2},$$

where $\rho = 2.20 \,\text{g cm}^{-3}$, $m = 60.1 \,\text{g mol}^{-1}$, $\langle b \rangle_{Si} = 4.15 \,\text{fm}$ (as above) and $\langle b \rangle_O = 5.80 \,\text{fm}$. The equivalent formula for X-rays is the same as eqn (4.11), except that the $\langle b \rangle_j$ are replaced by a product of the atomic number, Z_j, and the Thomson scattering length, r_e, which was met earlier in eqn (3.19):

$$\boxed{\langle b \rangle_j \rightarrow Z_j r_e} \,. \tag{4.12}$$

As such, β is proportional to the electron density. In the case of the silicon substrate and its oxide layer, the X-ray SLDs are 19.7 and $18.6 \times 10^{-6} \,\text{\AA}^{-2}$ respectively.

Although the Fourier relationship between the depth profile $\beta(z)$ and the reflectivity data $R(Q)$ is apparent from eqn (4.9), it is often easier to make the connection with the physical insight gained in Section 2.5 by using it in an alternative form involving the derivative $\mathrm{d}\beta/\mathrm{d}z$:

$$\boxed{R(Q) \approx \frac{16 \pi^2}{Q^4} \left| \int_{-\infty}^{\infty} \frac{\mathrm{d}\beta}{\mathrm{d}z} \, \mathrm{e}^{-izQ} \, \mathrm{d}z \right|^2} \,, \tag{4.13}$$

which follows from eqn (4.9) through 'integration by parts'. The mild technical proviso in the derivation, that $\beta \rightarrow 0$ as $z \rightarrow \pm\infty$, must be fulfilled inherently if $\beta(z)$ is to satisfy the Dirichlet condition.

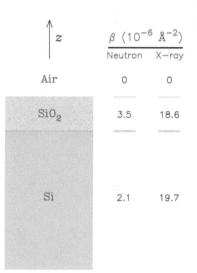

	β (10^{-6} Å^{-2})	
	Neutron	X-ray
Air	0	0
SiO$_2$	3.5	18.6
Si	2.1	19.7

4.1.1 Substrate only

As a first example of eqn (4.13) in action, let's consider the simplest possible situation: a bare substrate, with an SLD of β_s. For an ideal surface, the depth profile is a *Heaviside* function with a step discontinuity at the interface:

$$\beta(z) = \begin{cases} \beta_s & \text{for } z<0, \\ 0 & \text{for } z>0. \end{cases} \tag{4.14}$$

It's illustrated in Fig. 4.2 for the case of silicon with neutrons. Since the gradient of this $\beta(z)$ is zero everywhere except $z=0$, the derivative is a scaled δ-function,

$$\frac{\mathrm{d}\beta}{\mathrm{d}z} = -\beta_s\,\delta(z). \tag{4.15}$$

Being extremely narrow, its Fourier transform is uniform:

$$\beta_s \int\limits_{-\infty}^{\infty} \delta(z)\,\mathrm{e}^{-izQ}\,\mathrm{d}z = \beta_s\,\mathrm{e}^0 = \beta_s.$$

Hence, eqn (4.13) predicts a smooth fall off in reflectivity with the prefactor dependence of Q^{-4}:

$$R(Q) \approx \frac{16\,\pi^2\,\beta_s^2}{Q^4}. \tag{4.16}$$

Such a gradual decay is evident in Fig. 4.2, although the trend does not hold close to the origin. This break down is inevitable because eqn (4.16) violates the physical constraint that $R\leqslant 1$ as $Q\to 0$. We'll leave a discussion of the low-Q limitations of eqn (4.13) to Section 4.2, and continue exploring how a basic knowledge of Fourier transforms can help us to understand reflectivity data.

Equation (4.16) also predicts that the reflectivity will scale with the square of β_s. This dependence is verified by the vertical shift of

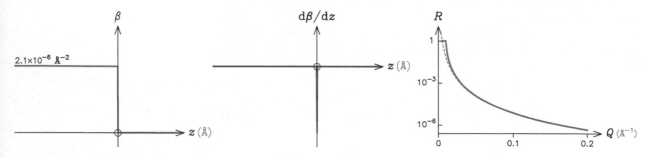

Fig. 4.2 The SLD depth profile, $\beta(z)$, and its derivative, $\mathrm{d}\beta/\mathrm{d}z$, for a bare substrate with a 'perfect' surface; the choice of β_s corresponds to silicon for neutron scattering. The reflectivity data predicted by eqn (4.16), indicated by the dashed grey line, is in good agreement with the 'true' $R(Q)$ for $Q>0.02\,\text{Å}^{-1}$.

$R(Q)$ in Fig. 4.3, which shows the situation when the SLD is about three times greater than that in Fig. 4.2.

So far, we have considered substrates with very sharp interfaces. The transition of the SLD at a boundary tends to be diffuse in practice, and an example is given with the dashed grey line in Fig. 4.3. This characteristic of the surface is often modelled by a Gaussian, so that the derivative

$$\frac{\mathrm{d}\beta}{\mathrm{d}z} = -\frac{\beta_{\mathrm{s}}}{\sigma\sqrt{2\pi}} \exp\left(-\frac{z^2}{2\,\sigma^2}\right), \qquad (4.17)$$

instead of the δ-function of eqn (4.15); the latter is recovered in the limit of $\sigma \to 0$. The Fourier transform is also Gaussian, but of the reciprocal width $1/\sigma$:

$$\int_{-\infty}^{\infty} \frac{\mathrm{d}\beta}{\mathrm{d}z}\, \mathrm{e}^{-\mathrm{i}zQ}\, \mathrm{d}z = -\beta_{\mathrm{s}} \exp\left(-\frac{\sigma^2 Q^2}{2}\right).$$

Substitution into eqn (4.13) predicts that the reflectivity will be that of eqn (4.16) modulated by a Gaussian with a standard deviation of $\left(\sigma\sqrt{2}\right)^{-1}$:

$$R(Q) \approx \frac{16\,\pi^2\,\beta_{\mathrm{s}}^2}{Q^4} \times \exp\left(-\sigma^2 Q^2\right). \qquad (4.18)$$

The more rapid decay with increasing Q expected from a bare substrate with a 'rough' surface can be seen in Fig. 4.3. In fact, this is a general feature of all SLD functions that contain diffuse interfaces and is akin to the Debye–Waller factor of Section 3.3.2. It is easiest to understand in detail when all the layer boundaries have the same (or similar) level of roughness, σ, through the use of the convolution theorem of Section 2.4.1,

$$R(Q) \approx R_{\mathrm{o}}(Q) \times \mathrm{e}^{-\sigma^2 Q^2}, \qquad (4.19)$$

where $R_{\mathrm{o}}(Q)$ is the equivalent reflectivity curve for a sample with sharp interfaces.

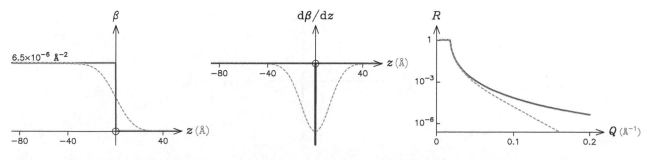

Fig. 4.3 The SLD depth profile, $\beta(z)$, its derivative, $\mathrm{d}\beta/\mathrm{d}z$, and the resultant reflectivity curve, $R(Q)$, for a bare substrate with β_{s} about three times greater than that in Fig. 4.2. The cases of both a sharp interface (blue) and a diffuse one (dashed grey), with a Gaussian roughness of $\sigma = 12\,\text{\AA}$, are shown.

4.1.2 One uniform layer

The next simplest situation involves a single uniform layer, of thickness L and SLD β_1, deposited on a substrate. The ideal depth profile has two step discontinuities,

$$\beta(z) = \begin{cases} \beta_s & \text{for } z < -L, \\ \beta_1 & -L < z < 0, \\ 0 & \text{for } z > 0, \end{cases} \qquad (4.20)$$

whose derivative is a pair of scaled δ-functions:

$$\frac{\mathrm{d}\beta}{\mathrm{d}z} = (\beta_1 - \beta_s)\,\delta(z+L) - \beta_1\,\delta(z). \qquad (4.21)$$

Three cases are shown in Fig. 4.4, all with a β_s of $2.1 \times 10^{-6}\,\text{Å}^{-2}$: the solid blue line has $L = 200\,\text{Å}$ and $\beta_1 = 3.5 \times 10^{-6}\,\text{Å}^{-2}$; the dotted line in light blue has a layer of the same thickness but a higher value of β_1, whereas the dashed grey line has the same SLD as the first but is only half as wide. The integral property of a δ-function, given in eqn (2.54), leads to

$$\int_{-\infty}^{\infty} \frac{\mathrm{d}\beta}{\mathrm{d}z}\, \mathrm{e}^{-\mathrm{i}zQ}\,\mathrm{d}z = (\beta_1 - \beta_s)\,\mathrm{e}^{\mathrm{i}LQ} - \beta_1.$$

The multiplication of the right-hand side with its complex conjugate, to obtain the modulus-squared of the Fourier transform of $\mathrm{d}\beta/\mathrm{d}z$, and the use of the trigonometric identity $\mathrm{e}^{\mathrm{i}LQ} + \mathrm{e}^{-\mathrm{i}LQ} = 2\cos(LQ)$, enables eqn (4.13) to be evaluated as

$$R(Q) \approx \frac{16\,\pi^2}{Q^4}\left[\beta_1^2 + (\beta_1 - \beta_s)^2 - 2\,\beta_1(\beta_1 - \beta_s)\cos(LQ)\right], \qquad (4.22)$$

where β_1 and β_s have implicitly been taken as real. Superimposed on the familiar Q^{-4} decay in the reflectivity, therefore, is a sinusoidal variation with a repeat Q-distance of

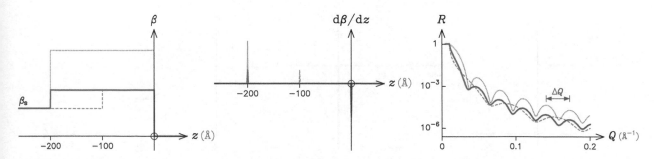

Fig. 4.4 The SLD depth profile, $\beta(z)$, its derivative, $\mathrm{d}\beta/\mathrm{d}z$, and the resultant reflectivity curve, $R(Q)$, for a single uniform layer deposited on a substrate. Three cases are shown, all with the same β_s and sharp interfaces: the one in dotted light blue has a higher value of β_1 than the solid blue line, whereas the dashed grey one is narrower.

$$\Delta Q = 2\pi/L. \tag{4.23}$$

This can seen in Fig. 4.4, where ΔQ is twice as large for the case of the narrower layer in the dashed grey line.

The dependence of the amplitude, $A = 2\beta_1(\beta_1 - \beta_s)$, and the offset, $B = \beta_1^2 + (\beta_1 - \beta_s)^2$, of the oscillations, $B - A\cos(LQ)$, on the magnitude of β_1 and its difference from β_s can also be verified in Fig. 4.4: the visibility of the fringes, and the overall level of the reflectivity, is comparable for the solid blue and the dashed grey cases, but is noticeably greater for the dotted light blue one. As required theoretically, the expression of eqn (4.22) reduces to eqn (4.16) when either $\beta_1 = 0$ or $\beta_1 = \beta_s$.

The last thing to note is that the fringes start to become washed out as Q increases. This is not predicted by the analysis outlined here, but was encoded in the computer simulation of the data to mimic the blurring effect of a finite experimental resolution. This *point-spread function* was approximated by a Gaussian whose standard deviation, σ, varied linearly with Q; a fractional value of $\sigma/Q = 2\%$ was used, although 3 to 5 % is more typical in practice. Such a resolution blurring was also included in Figs. 4.2 and 4.3, but is imperceptible because the reflectivity curves of eqns (4.16) and (4.18) do not contain any fine structure. Statistical noise, which limits the smallest reflectivity that can be measured reliably, was omitted to enhance clarity.

4.1.3 A short multilayer

The next example, shown in Fig. 4.5, involves three alternating layers of two materials deposited on a substrate; for simplicity, all the thicknesses have been made the same. Rather than relating the corresponding reflectivity curve to the SLD depth profile analytically, let's try to understand it qualitatively with the help of the physical insight into Fourier transforms gained in Section 2.5. Since $d\beta/dz$ for the multilayer is reminiscent of the aperture function, $A(x)$, in

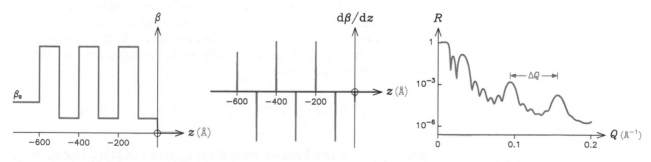

Fig. 4.5 The SLD depth profile, its derivative and the resultant reflectivity curve for a short multilayer deposited on a substrate. The reciprocal of the distance between the prominent peaks in $R(Q)$, $2\pi/\Delta Q$, gives the repeating distance of 100 Å in $\beta(z)$, whereas the separation of the finer fringes indicates the overall thickness of the sample.

Fig. 2.17, at least when the minus sign of half the δ-function amplitudes is ignored, we might expect the modulus-squared of its Fourier transform to exhibit structure similar to the diffraction intensity, $\mathrm{I}(q)$, for a finite grating. This is indeed the case, with $R(Q)$ showing a regular pattern of prominent peaks interspersed with a number of finer fringes (superimposed on the Q^{-4} decay); the latter become washed out at higher Qs due to resolution blurring. The reciprocal of the spacing between the main peaks, $2\pi/\Delta Q$, gives the repeat distance of 100Å in $\beta(z)$. The number of finer fringes and their spacing, as well as the width of the major peaks, is related to the overall thickness of the multilayer.

The sort of intuitive Fourier reasoning used above can also be applied to other situations. The pair of δ-functions defining $\mathrm{d}\beta/\mathrm{d}z$ in Fig. 4.4, for example, resemble the aperture function in a Young's double slit setup. By analogy with Fig. 2.14, therefore, we would expect the reflectivity curve to contain uniform cosine fringes whose spacing depends inversely on the thickness of the associated single layer. These types of argument demonstrate the power of the Fourier relationship of eqn (4.13) in helping us to understand the connection between reflectivity measurements and the compositional depth structure of the sample. Before discussing its breakdown at low Q, let's highlight the difficulty inherent in Fourier analysis when the complex phase is not available.

4.1.4   A phaseless Fourier ambiguity

Figure 4.6 shows two hypothetical samples of the same total thickness, consisting of one and two layers, deposited on a substrate with $\beta_{\mathrm{s}} = 2.1 \times 10^{-6}$ Å^{-2}. Even though the SLD depth profiles are distinctly different, the resultant reflectivity curves are virtually identical; the only slight discrepancy is at the first minimum. This ambiguity is an example of the information lost in diffraction experiments due to the inability of measuring the phase of the scattered wave: eqns (2.58), (3.32), (3.33) and (4.13) all involve a modulus-squared in the final step.

The reason for the similarity of the reflectivity curves in Fig. 4.6 can be understood by considering the auto-correlation functions of the relevant $\mathrm{d}\beta/\mathrm{d}z$. As discussed in Section 2.4.2, the ACF gives a real-space representation of the information contained in the intensity of a diffraction pattern:

$$\mathrm{ACF}(z) \;=\; \int_{-\infty}^{\infty} \mathrm{f}(t)^* \, \mathrm{f}(z{+}t) \, \mathrm{d}t \;=\; \int_{-\infty}^{\infty} |\mathrm{F}(Q)|^2 \, \mathrm{e}^{\mathrm{i}zQ} \, \mathrm{d}Q \,,$$

where $\mathrm{F}(Q)$ is the Fourier transform of $\mathrm{f}(z)$. In the present case, $\mathrm{f}(z) = \mathrm{d}\beta/\mathrm{d}z$ is purely real and comprises of two or three δ-functions. As such, it's easiest to calculate the ACF by direct inspection. There is always a large positive peak at the origin because, at zero separa-

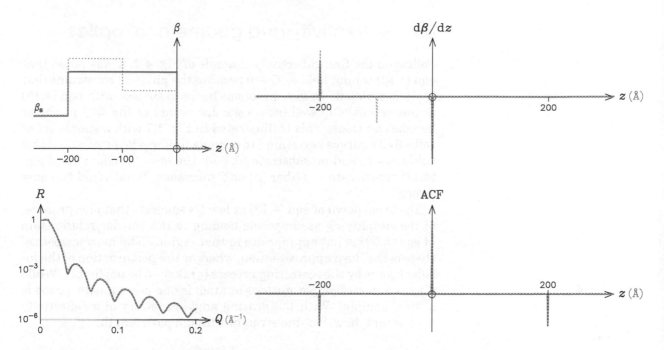

Fig. 4.6 The SLD depth profile, $\beta(z)$, of two hypothetical samples, plotted in solid blue and dotted grey, deposited on the same substrate, and the resultant reflectivity curves, $R(Q)$. The derivatives of the profiles, $\mathrm{d}\beta/\mathrm{d}z$, and their auto-correlation functions, ACF, are also shown.

tion, everything correlates with itself. The negative spikes at $\pm 200\,\text{Å}$ in the ACF arise from the interference between the front and back of the sample, or the pair of δ-functions with amplitudes of opposite signs at $z = 0$ and $z = -200\,\text{Å}$ in $\mathrm{d}\beta/\mathrm{d}z$. The extra components that might be expected in the ACF for the two-layer sample, at $\pm 100\,\text{Å}$, are absent because there is a complete cancellation between a positive and a negative contribution from two pairs of δ-functions in $\mathrm{d}\beta/\mathrm{d}z$ at that separation. This equality of the ACFs explains the similarity of the reflectivity curves. They would be identical if eqn (4.13) held exactly for all Q!

The purpose of this simple example was to illustrate how difficult it is to infer uniquely the nature of an SLD function solely from a set of diffraction data. Since the problem of ambiguities will only be exacerbated with complexity in $\beta(z)$, the reflectivity measurements must be supplemented with additional information. Perhaps the number of layers is known from the method of sample preparation, or some of the theoretical alternatives do not satisfy physical or chemical constraints. In the case of Fig. 4.6, the two-layer model may be suspect due to the fortuitous symmetries in the SLD depth profile. A reliance on such prior knowledge, assumptions or beliefs can sometimes be avoided through clever experiments, but most of these require multiple sets of data connected by carefully controlled changes.

4.2 Reflectivity and geometrical optics

Following the first reflectivity example of Fig. 4.2, it was noted that eqn (4.13) cannot hold as $Q \to 0$ because the physical constraint that $R \leqslant 1$ is violated. In fact, problems begin to surface with eqn (4.13) at low values of Q well before the divergence of the Q^{-4} prefactor becomes an issue. This is illustrated in Fig. 4.7 with a simple set of reflectivity curves pertaining to a single uniform layer, of SLD β and thickness T, and no substrate ($\beta_s = 0$): the low-Q predictions of eqn (4.13) deteriorate as either $|\beta|$ or T increases. What could be going wrong?

The breakdown of eqn (4.13) at low Qs suggests that one, or more, of the simplifying assumptions leading to the Fourier relationship of eqn (3.33) is not appropriate in that regime. The most suspect of these is the Born approximation, whereby the perturbation of the incident wave by the scattering process is taken to be negligible. While a weak interaction per nucleus or atom is one prerequisite, so too is a 'thin' sample. With the grazing angle geometry of a reflectivity experiment, however, the straight through path length,

$$L = T/\sin\theta \, ,$$

can be rather long. We should not be too surprised, therefore, if the validity of eqn (4.13) becomes doubtful as θ decreases, especially when T is large or the layers consist of strong scatterers. Although the inverse dependence of L on $\sin\theta$ explains why the preceding Fourier analysis might run into difficulties as $Q \to 0$ at a fixed wavelength, it offers no insight on its demise at longer λs. The answer to that conundrum probably has to do with the reciprocal relationship between the momentum of an incident particle and its wavelength: the more energetic the projectile, the less it's perturbed.

To obtain an accurate description of the reflectivity curve at all values of Q, we must adopt a different approach to the problem. It's sometimes called the *dynamical* theory of scattering, as opposed to the kinematical one of the previous section.

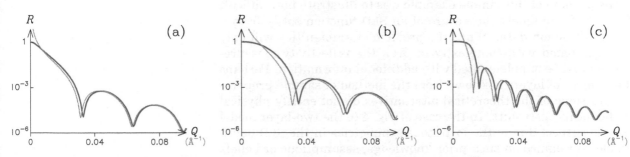

Fig. 4.7 The deviation of the kinematical approximation (dotted grey line) to the reflectivity curve (in solid blue), at low values of Q, for a uniform layer of SLD β and thickness T (with $\beta_s = 0$). In (a), $\beta = 2.0 \times 10^{-6}\,\text{Å}^{-2}$ and $T = 200\,\text{Å}$; (b) and (c) differ from this by having a larger SLD, of $6.0 \times 10^{-6}\,\text{Å}^{-2}$, or being thicker, at $T = 500\,\text{Å}$, respectively.

4.2.1 Refractive index

The reflection of light from planar surfaces has been studied for centuries, and is intimately linked with *refraction* at the boundary: part of the intensity bounces off at the angle of incidence, θ_i, while the rest is transmitted with a change in direction, to θ_t. This accounts for the appearance of the kink in objects standing in a glass of water. For a given material below the air–matter interface, the ratio of the cosines of these angles is found to be a constant,

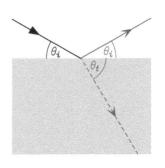

$$n = \frac{\cos\theta_i}{\cos\theta_t}, \tag{4.24}$$

called the *refractive index*. The value of n can depend on the incident wavelength, λ_i, which explains why white (polychromatic) light is split into a rainbow by a prism.

Although the refractive law of eqn (4.24) is usually attributed to Snell (1621) or Descartes (1637), it is described in the earlier work of Ibn Sahl (984).[†] Likewise, a rudimentary form of Fermat's *principle of least time* (1662) used to derive this observation theoretically can be found in the writings of Ibn al-Haytham (1021) or, even further back, in the ideas of Hero of Alexandria (60).[‡] At its heart lies the fact that the speed of light varies with the medium through which it travels. In conjunction with Huygens' wave postulate (1678), the refractive index is then seen to be a ratio of speeds or wavelengths as much as the cosines of angles:

[†] Kwan *et al.* (2002), Rashed (1990), Wolf (1995).

[‡] Mihas (2005)

$$n = \frac{c_i}{c_t} = \frac{\lambda_i}{\lambda_t}, \tag{4.25}$$

$$c = \nu\lambda$$

where the frequency of the (electromagnetic) oscillations, ν, remains unchanged across the boundary. Let's see how the refractive index and the mathematics of one-dimensional waves yield the reflectivity curve for a bare substrate that is valid for all Q.

4.2.2 A single boundary

Perpendicular to the planar interfacial surface ($z = 0$), the incident wave takes the form

$$\psi_i(z) = \psi_0 \, e^{-\mathrm{i} z k_i \sin\theta}, \tag{4.26}$$

where $k_i = 2\pi/\lambda$ and $\theta_i = \theta$ for consistency with Fig. 4.1. Part of this signal is reflected at the boundary, and travels in the $+z$ direction with a complex amplitude r,

$$\psi_r(z) = r \, e^{\mathrm{i} z k_i \sin\theta}, \tag{4.27}$$

while the rest, quantified by t, is transmitted:

$$\psi_t(z) = t \, e^{-\mathrm{i} z k_t \sin\theta_t}. \tag{4.28}$$

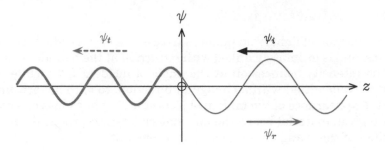

Fig. 4.8 The net displacement of a taut string, ψ, when an incident wave, ψ_i, encounters a sudden change in the mass per unit length. A portion, ψ_r, of the incoming signal is reflected at the boundary (at $z=0$) while the rest, ψ_t, is transmitted.

The wavenumber k_t and the sine of θ_t are related to the incoming values through the refractive index n:

$$\sin^2\theta + \cos^2\theta \equiv 1$$

$$k_t = \frac{2\pi n}{\lambda} \quad \text{and} \quad 1 - \sin^2\theta_t = \frac{1 - \sin^2\theta}{n^2}.$$

The situation is like that of a wave on a taut string when it encounters a sudden change in the mass per unit length, and is illustrated in Fig. 4.8. Drawing on this analogy, constraints on r and t can be obtained by imposing the boundary conditions that both the fluctuation, ψ, and its derivative, $\mathrm{d}\psi/\mathrm{d}z$, must match up at the transition. That is to say, at $z=0$,

$$\psi_i + \psi_r = \psi_t \quad \text{and} \quad \frac{\mathrm{d}}{\mathrm{d}z}(\psi_i + \psi_r) = \frac{\mathrm{d}\psi_t}{\mathrm{d}z}. \tag{4.29}$$

Substitution from eqns (4.26)–(4.28) then leads to the simultaneous equations

$$\frac{\mathrm{d}}{\mathrm{d}z}\left(\mathrm{e}^{Kz}\right) = K\,\mathrm{e}^{Kz}$$

$$\psi_{\mathrm{o}} + r = t \quad \text{and} \quad (\psi_{\mathrm{o}} - r)\,k_i \sin\theta = t\,k_t \sin\theta_t,$$

which can easily be solved to yield

$$t = \frac{2\psi_{\mathrm{o}}}{(1+\alpha)} \quad \text{and} \quad r = \left(\frac{1-\alpha}{1+\alpha}\right)\psi_{\mathrm{o}} \tag{4.30}$$

where

$$\alpha = \frac{k_t \sin\theta_t}{k_i \sin\theta} = \frac{n}{\sin\theta}\left[1 - \frac{(1-\sin^2\theta)}{n^2}\right]^{1/2}. \tag{4.31}$$

If the term in square brackets is negative, α will be an imaginary number and the reflectivity,

$$\alpha^2 \leqslant 0 \implies \mathcal{R}e\{\alpha\} = 0$$
$$\implies R = 1$$

$$R = \left|\frac{r}{\psi_{\mathrm{o}}}\right|^2 = \frac{1 - 2\,\mathcal{R}e\{\alpha\} + |\alpha|^2}{1 + 2\,\mathcal{R}e\{\alpha\} + |\alpha|^2}, \tag{4.32}$$

will be unity (just as for $\alpha = 0$). As this requires

$$\sin^2\theta \leqslant 1 - n^2,$$

complete reflection occurs from materials with $n < 1$ for glancing angles below a critical value of

$$\theta_c = \sin^{-1}\left(\sqrt{1-n^2}\right). \tag{4.33}$$

The formulae for $|r/\psi_0|^2$ and $|t/\psi_0|^2$ resulting from eqns (4.30) and (4.31) are known as *Fresnel equations*.

Having seen how the low-Q behaviour of the reflectivity curve is correctly predicted by this formulation, let's make sure that its decay agrees with eqn (4.16) at large value of Q. In fact, this necessary condition will enable us to work out the relationship between the SLD, β, the wavelength, λ, and the refractive index, n. Rewriting eqn (4.31) for α in terms of Q with eqn (4.1), and using the binomial expansion for the square root,

$$\alpha \approx 1 - \frac{8\pi^2\left(1-n^2\right)}{\lambda^2\,Q^2}\,, \qquad\qquad \sqrt{1-\epsilon} = 1 - \frac{\epsilon}{2} - \frac{\epsilon^2}{8} - \cdots$$

which becomes ever more accurate as $Q \to \infty$. Substitution into eqn (4.32), with $1 + \alpha \approx 2$, leads to

$$R = \left|\frac{1-\alpha}{1+\alpha}\right|^2 \approx \frac{16\,\pi^2}{Q^4}\left[\frac{\pi\left(1-n^2\right)}{\lambda^2}\right]^2.$$

A comparison with eqn (4.16) shows that the term in square brackets must be equated with β, so that

$$n^2 \approx 1 - \frac{\beta\lambda^2}{\pi}\,. \tag{4.34}$$

In conjunction with eqns (4.1) and (4.33), therefore, θ_c can be translated into an equivalent critical Q,

$$Q_c = 4\sqrt{\pi\beta}\,, \tag{4.35}$$

below which $R = 1$ for $\beta > 0$; complete reflection cannot occur if the SLD is negative. Finally, the square root of eqn (4.34) with the binomial expansion yields

$$\boxed{n \approx 1 - \frac{\beta\lambda^2}{2\pi}}\,, \tag{4.36}$$

which is the dependence of the refractive index on the wavelength and scattering-length density derived by Fermi (1950) through a consideration of the signal transmitted by a planar slab of uniform material.

4.2.3   Multiple boundaries

The analysis for the reflection at a single boundary can be generalized to the case of multiple transitions in the refractive index. The situation for N layers is illustrated schematically in Fig. 4.9, with the SLD depth profile being defined by

$$\beta_{N+1} = \beta_{\mathrm{s}} \quad \text{and} \quad \beta_0 = 0$$

$$\beta(z) = \begin{cases} \beta_{\mathrm{s}} & \text{for } z < z_N, \\ \beta_j & \text{for } z_j < z < z_{j-1}, \\ 0 & \text{for } z > 0, \end{cases} \tag{4.37}$$

where $j = 1, 2, 3, \ldots, N$ and $z_0 = 0$. The displacement ψ of the oscillation in the uniform region j can always be expressed as

$$\psi_0(z) = \psi_r(z) + \psi_i(z)$$

$$\psi_j(z) = A_j \, \mathrm{e}^{\mathrm{i}z\,k_j} + B_j \, \mathrm{e}^{-\mathrm{i}z\,k_j}, \tag{4.38}$$

where A_j and B_j are the (complex) amplitudes of waves travelling in the positive and negative z directions, ψ_{j+} and ψ_{j-} respectively, with a wavenumber

$$k_j = \left(\frac{2\pi}{\lambda_j}\right) \sin\theta_j = \left(\frac{2\pi n_j}{\lambda}\right) \sqrt{1 - \frac{1 - \sin^2\theta}{n_j^2}},$$

where λ and θ are the wavelength and glancing angle of the incident beam, and n_j is the refractive index of the jth layer. A rearrangement of the right-hand side, including the substitution for n_j^2 from eqn (4.34) and $\sin\theta/\lambda$ from eqn (4.1), allows the relevant wavenumber to be written as

$$2\,k_j = \sqrt{Q^2 - 16\pi\beta_j}\,. \tag{4.39}$$

Given the transmission condition that

$$\psi_{N+1}(z) = \psi_t(z)$$

$$A_{N+1} = 0\,, \tag{4.40}$$

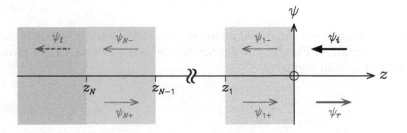

Fig. 4.9 The oscillatory displacement ψ in the uniform region j, $\psi_j(z)$, can be expressed as the sum of a forward- and a backward-moving wave, ψ_{j+} and ψ_{j-}, with a wavenumber, k_j, determined by the incident wavelength, λ, the glancing angle, θ, and the refractive index, n_j.

because there is no reverse wave in the substrate, our task is to match $\psi(z)$ and its derivative, $\psi'(z)$, at each of the refractive index boundaries,

$$\psi_{j+1}(z_j) = \psi_j(z_j) \quad \text{and} \quad \psi'_{j+1}(z_j) = \psi'_j(z_j) \qquad (4.41)$$

$$\psi'(z) = \frac{\mathrm{d}\psi}{\mathrm{d}z}$$

for $j = 0, 1, 2, \ldots, N$, with the aim of ascertaining $|A_0/B_0|^2$ to obtain the reflectivity.

Since eqn (4.38) and its derivative relate $\psi_j(z)$ and $\psi'_j(z)$, between z_j and z_{j-1}, to A_j and B_j,

$$A_j = \frac{\mathrm{i}\,k_j\,\psi_j(z) + \psi'_j(z)}{2\mathrm{i}\,k_j\,\mathrm{e}^{\mathrm{i}z\,k_j}} \quad \text{and} \quad B_j = \frac{\mathrm{i}\,k_j\,\psi_j(z) - \psi'_j(z)}{2\mathrm{i}\,k_j\,\mathrm{e}^{-\mathrm{i}z\,k_j}},$$

the reflectivity is given by

$$\boxed{R(Q) = \left| \frac{\mathrm{i}\,k_0\,\psi(0) + \psi'(0)}{\mathrm{i}\,k_0\,\psi(0) - \psi'(0)} \right|^2} \qquad (4.42)$$

$$k_0 = Q/2$$

and eqn (4.40) implies that

$$\psi'(z_N) = -\mathrm{i}\,k_{\mathrm{s}}\,\psi(z_N). \qquad (4.43)$$

$$k_{\mathrm{s}} = \sqrt{k_0^2 - 4\pi\beta_{\mathrm{s}}}$$

Starting from an assignment of ψ and ψ' at $z = z_N$ which is consistent with eqn (4.43), such as

$$\psi(z_N) = 1 \quad \text{and} \quad \psi'(z_N) = -\mathrm{i}\,k_{\mathrm{s}},$$

the wave of eqn (4.38) and its derivative can be propagated through the Nth uniform layer, of thickness T_N, to the other boundary at $z = z_{N-1}$:

$$\psi(z_{j-1}) = \cos(k_j T_j)\,\psi(z_j) + \frac{\sin(k_j T_j)}{k_j}\,\psi'(z_j), \qquad (4.44)$$

$$k_j = \sqrt{k_0^2 - 4\pi\beta_j}$$

$$\psi'(z_{j-1}) = -k_j \sin(k_j T_j)\,\psi(z_j) + \cos(k_j T_j)\,\psi'(z_j), \qquad (4.45)$$

$$T_j = z_{j-1} - z_j$$

with $j = N$. The continuity conditions of eqn (4.41) mean that eqns (4.44) and (4.45) can be used recursively until ψ and ψ' are evaluated at the surface of the sample, at $z = z_0 = 0$, whence eqn (4.42) yields the reflectivity.

$$\begin{pmatrix} \psi \\ \psi' \end{pmatrix}_{z_N} \rightarrow \begin{pmatrix} \psi \\ \psi' \end{pmatrix}_{z_{N-1}} \rightarrow \begin{pmatrix} \psi \\ \psi' \end{pmatrix}_{z_{N-2}} \rightarrow \cdots$$

This type of iterative procedure for calculating $R(Q)$ was first described by Abelès (1948) and Parratt (1954). Although it gives an accurate prediction of the reflectivity for all values of Q, the results lack the physical intuition allowed by the Fourier analysis of the kinematic approximation. Before moving on to the next example of elastic scattering, in Chapter 5, let us briefly compare the strengths and weaknesses of X-rays and neutrons; we'll also touch upon alternative techniques for surface studies.

From eqn (4.45),

$$\psi'(z+dz) = \psi'(z) \underbrace{- k^2(z)\,\psi(z)\,dz}_{+\dfrac{d\psi'}{dz}}$$

[†] W. A. Hamilton (1989), private communication.

$$2\,k(z)\,\frac{dk}{dz} = 4\pi\,\frac{d\beta}{dz}$$

The kinematic approximation to dynamical reflectivity

For an arbitrary SLD profile, $\beta(z)$, described by a huge number ($N \to \infty$) of extremely thin layers (of width dz), the problem becomes one of solving the second-order *differential equation*

$$\frac{d^2\psi}{dz^2} + \left[\frac{Q^2}{4} - 4\pi\,\beta(z)\right]\psi(z) = 0, \qquad (4.46)$$

subject to pure transmission within the substrate ($z < z_s$), to evaluate eqn (4.42) for the reflectivity. While an analytical solution for $\psi(z)$ is difficult to obtain in general, progress can be made when $Q^2 \gg 16\,\pi\,\beta(z)$; it leads to the kinematical formulae of Section 4.1. A derivation of eqn (4.13) is outlined below.[†]

Consider the function

$$r(z) = \frac{i\,k(z)\,\psi(z) + \psi'(z)}{i\,k(z)\,\psi(z) - \psi'(z)},$$

where $k^2(z) = Q^2/4 - 4\pi\,\beta(z)$, so that the reflectivity of eqn (4.42) is equal to $|r(0)|^2$. In conjunction with eqn (4.46), and a fair amount of effort, it can be shown to satisfy the first-order differential equation

$$\frac{dr}{dz} - i\,2k(z)\,r = \frac{\pi}{k^2(z)}\,\frac{d\beta}{dz}\,(r^2 - 1).$$

This becomes tractable in the limit of large Q and low reflectance, when $k(z) \to Q/2$ and $|r| \ll 1$, because it can then be solved with the aid of an *integrating factor*:

$$\frac{d}{dz}\left(r\,e^{-izQ}\right) \approx -\frac{4\pi}{Q^2}\,\frac{d\beta}{dz}\,e^{-izQ}.$$

Integrating both sides with respect to z from within the substrate, where $r=0$, to the surface of the sample, at $z=0$,

$$\left[r\,e^{-izQ}\right]_{-\infty}^{0} = r(0) \approx -\frac{4\pi}{Q^2}\int_{-\infty}^{0}\frac{d\beta}{dz}\,e^{-izQ}\,dz.$$

The kinematic formula of eqn (4.13) follows from the observation that the upper limit of the integral can be set to $+\infty$ since $d\beta/dz = 0$ for $z > 0$.

4.3 X-rays, neutrons and other techniques

The differences between X-rays and neutrons in terms of reflectivity measurements are much the same as for all scattering experiments, and follow from the properties of these probes and their production facilities. Due to their accessibility, laboratory-based X-ray sources are usually the first port-of-call; reflectivities as low as 10^{-9}, and Qs as high as an Å^{-1}, are achievable under favourable conditions. By

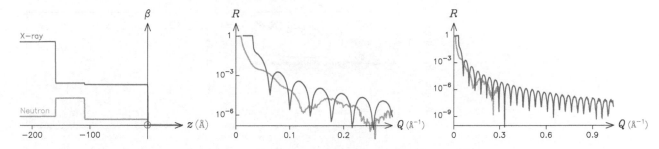

Fig. 4.10 The X-ray and neutron SLD depth profiles, $\beta(z)$, and the corresponding reflectivity curves, $R(Q)$, for the same sample. While the former has a better signal-to-noise ratio, and is reliable out to a much higher value of Q, the lack of contrast in the X-ray SLD gives rise to uniform fringes that only indicate the overall thickness of the sample (160Å). This length-scale can also be seen in the poorer quality neutron data, but the additional broader variation marks the presence of the (50Å) subdivision.

comparison, it's difficult to go below $R \approx 10^{-6}$, or beyond $Q \approx 0.2\text{Å}^{-1}$, with neutrons because of the combination of a poorer incident flux and a greater background signal from incoherent scattering. These drawbacks can be offset by the increased contrast in the SLD depth profile for neutrons, leading to a richer structure in the data, since the scattering lengths do not vary monotonically with atomic number. Indeed, the ability to alter $\beta(z)$ through isotopic substitution, or with the polarization of the incident beam for magnetic samples, makes neutrons a versatile probe.

A complementary technique for ascertaining the refractive index depth profile is provided by *ellipsometry*, whereby the reflectivities parallel and perpendicular to an incident plane-polarized light beam are analyzed simultaneously. Atomic-scale detail can be inferred despite the much longer wavelengths, but thin samples are troublesome. Information can be obtained in a more direct fashion with the bombardment of the sample with energetic ions, typically hydrogen or helium of several MeV, and an examination of the ejected 'debris'; such an *ion beam analysis* tends to have a depth resolution of only 10 to 20nm, however, and is restricted to work with solids. The measurement of the *surface tension* at a liquid surface can also be useful.

The main assumption made in this chapter was that of a layered sample which is invariant in the x–y plane; this reduced the analysis to a one-dimensional problem. If the approximation of eqn (4.2) is not adequate, a significant fraction of the scattered intensity will lie outside the narrow region defined by eqn (4.5). In terms of Fig. 4.1, there will be a substantial signal even when the angle of incidence, θ_i, is not equal to that of reflection, θ_r, or when they are not in the same plane. The *off-specular* pattern tends not to be highly structured, however, due to the lack of a regular repeating pattern in the x–y domain. Nevertheless, the spread of the *diffuse* intensity gives an indication of the characteristic length scale(s) in the plane perpendicular to z via their reciprocal (Fourier) relationship. If the sample is solid and the x–y correlations are down to features on

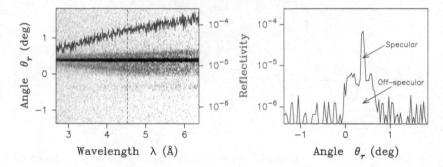

Fig. 4.11 A grey-scale image of the reflectivity, as a function of the neutron wavelength λ and reflection angle θ_r, from a monolayer of two phospholipid components on top of a water substrate (Hughes *et al.*, 2007). The blue lines show the corresponding one-dimensional data along the 'specular ridge', $\theta_r = \theta_i$ (the angle of incidence), and at a constant $\lambda \approx 4.5\text{Å}$.

the exposed surface (at $z \approx 0$), the origin of the off-specular signal can be explored with atomic force microscopy (AFM). *Brewster-angle microscopy* can help with the study of inhomogeneities on a liquid surface, but the spatial resolution is far from being atomic (closer to a μm than an Å).

Small-angle scattering and the big picture

5

Many biological and physical processes are controlled by the gross characteristics of large molecules and aggregates. Information about the size, shape and assemblies of these entities is obtained with small-angle scattering.

5.1 Diffraction and length scales

In Section 3.3, we learnt that the differential cross-section for elastic scattering, $(\mathrm{d}\sigma/\mathrm{d}\Omega)_{\mathrm{el}}$, is related to the SLD function, $\beta(\mathbf{r})$, through its Fourier transform:

$$\left(\frac{\mathrm{d}\sigma}{\mathrm{d}\Omega}\right)_{\mathrm{el}} \propto \left| \iiint_V \beta(\mathbf{r})\, \mathrm{e}^{\mathrm{i}\mathbf{Q}\cdot\mathbf{r}}\, \mathrm{d}^3\mathbf{r} \right|^2 , \qquad (5.1)$$

where V is the volume of the sample within which the scattering can occur. One of the most basic properties of Fourier transforms is the inverse relationship between length scales in real, $\mathbf{r} \equiv (x, y, z)$, and reciprocal, $\mathbf{Q} \equiv (Q_x, Q_y, Q_z)$, space. If the size of a compact object is defined by the three r-lengths d_x, d_y and d_z, for example, then the magnitude of its Fourier transform will decay towards zero beyond a Q-distance of around $2\pi/d_x$, $2\pi/d_y$ and $2\pi/d_z$ from the origin. Conversely, the lack of (reliable) scattering measurements for $|\mathbf{Q}|$-values larger than Q_{max} limits the level of the fine detail that can be inferred about the sample to of order $2\pi/Q_{\mathrm{max}}$. Experiments focusing on $|\mathbf{Q}| \lesssim 1\,\mathrm{nm}^{-1}$ provide low-resolution structural information and are, therefore, good for seeing the 'big picture': the overall configuration and assemblies of macromolecules, and other objects of a similar size. The scope and context of the technique is summarized in Fig. 5.1.

The relationship between the wavelength of the incident particle, the scattering angle and the modulus of the elastic wavevector transfer was given in Section 3.1.1 as

$$Q = \frac{4\pi \sin\theta}{\lambda} . \qquad (5.2)$$

Thus low-Q experiments benefit from large λs and entail measurements at small θs. The latter, achieved by having a long secondary

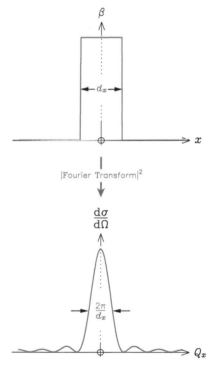

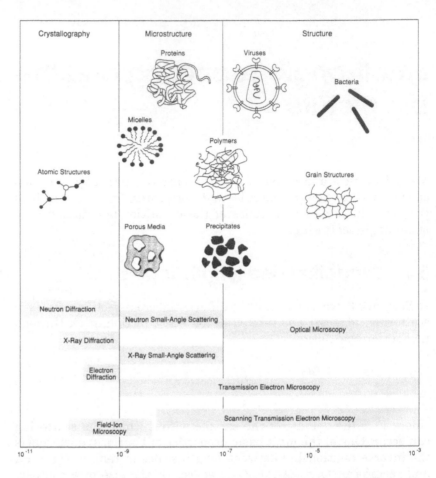

Fig. 5.1 An overview, following Pynn (1990), of how objects spanning length scales from 10^{-10} to 10^{-4} m can be probed by a variety of techniques. Small-angle X-ray and neutron scattering experiments are suited to low resolution structural studies of macromolecular complexes, porous materials and microscopic aggregates.

flight-path, as indicated in Fig. 5.2, gives rise to the general name of **S**mall-**A**ngle **S**cattering; it's often abbreviated to SAXS and SANS for X-rays and neutrons, respectively. By contrast, diffraction experiments requiring atomic resolution need data out to higher Qs and, hence, wider angles.

5.2   Size, shape and molecular form factors

Most small-angle scattering experiments are carried out on dilute solutions, so that the sample consists of isolated and isotropically oriented examples of the object of interest located randomly within a sea of solvent. If all N specimens are identical apart from their position, $\mathbf{r}_j$, and orientation, $\boldsymbol{\Theta}_j$, where $j = 1, 2, 3, \ldots, N$, then $\beta(\mathbf{r})$

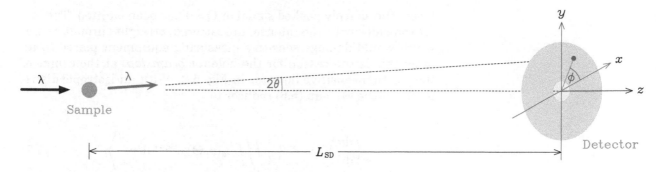

Fig. 5.2 The setup for a low-Q experiment requires a long sample-to-detector distance, L_{SD}, encased in an evacuated tank to avoid losses due to scattering by air molecules. The elastic wavevector transfer, $\mathbf{Q}$, is related to λ, θ and ϕ through eqn (3.8), with a signal at (x, y) corresponding to $\tan\phi = y/x$ and $\tan 2\theta = \sqrt{x^2+y^2}/L_{SD}$.

can be expressed as

$$\beta(\mathbf{r}) = \beta_o + \sum_{j=1}^{N} \widehat{\beta}_c(\mathbf{r}-\mathbf{r}_j, \boldsymbol{\Theta}_j), \qquad (5.3)$$

where β_o is the SLD of the solvent and

$$\widehat{\beta}_c(\mathbf{r}, \boldsymbol{\Theta}) = \begin{cases} \beta_c(\mathbf{r}, \boldsymbol{\Theta}) - \beta_o & \text{for } \mathbf{r} \text{ inside solute}, \\ 0 & \text{for } \mathbf{r} \text{ in the solvent}, \end{cases} \qquad (5.4)$$

where $\beta_c(\mathbf{r}, \boldsymbol{\Theta})$ is a compact SLD function which characterizes the entity being studied (centred at the origin and with orientation $\boldsymbol{\Theta}$); it is illustrated schematically in Fig. 5.3. For the simple case of a spherical complex, which has no special directions, with a constant SLD of β_1 and radius R,

$$\beta_c(\mathbf{r}, \boldsymbol{\Theta}) = \begin{cases} \beta_1 & \text{for } |\mathbf{r}| < R, \\ 0 & \text{otherwise}. \end{cases} \qquad (5.5)$$

Since the Fourier transform of an invariant is a δ-function, and those of shifted functions are related through a phase factor, as listed in Section 2.7, eqns (5.1) and (5.3) lead to

$$\left(\frac{d\sigma}{d\Omega}\right)_{el} \propto \left| (2\pi)^3 \, \delta(\mathbf{Q}) + \sum_{j=1}^{N} e^{i\mathbf{Q}\bullet\mathbf{r}_j} \iiint_{V_{\boldsymbol{\Theta}_j}} \widehat{\beta}_c(\mathbf{r}, \boldsymbol{\Theta}_j) \, e^{i\mathbf{Q}\bullet\mathbf{r}} \, d^3\mathbf{r} \right|^2,$$

where $V_{\boldsymbol{\Theta}}$ is the volume of space encapsulated by a solute structure that is centred at the origin with orientation $\boldsymbol{\Theta}$. When the summation is expanded and the modulus-squared evaluated, the part that adds up coherently, for uncorrelated positions $\mathbf{r}_j$ and $\mathbf{r}_{j'}$, is

$$\left(\frac{d\sigma}{d\Omega}\right)_{el} \propto \sum_{j=1}^{N} \left| \iiint_{V_{\boldsymbol{\Theta}_j}} \widehat{\beta}_c(\mathbf{r}, \boldsymbol{\Theta}_j) \, e^{i\mathbf{Q}\bullet\mathbf{r}} \, d^3\mathbf{r} \right|^2, \qquad (5.6)$$

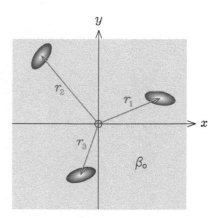

Fig. 5.3 A schematic illustration of eqn (5.3), with grey indicating solvent.

where the sharply peaked signal at $Q=0$ has been omitted. The latter coincides with the intense, unscattered, straight-through beam, which would damage sensitive measuring equipment placed in its path; this is the reason for the hole (or *beamstop*) at the centre of the position-sensitive detector in Fig. 5.2. With an isotropic distribution of the Θ_j, eqn (5.6) reduces to

$$\left(\frac{d\sigma}{d\Omega}\right)_{el} \propto N \left\langle \left| \iiint_{V_\Theta} \widehat{\beta}_c(\mathbf{r},\mathbf{\Theta})\, e^{i\mathbf{Q}\cdot\mathbf{r}}\, d^3\mathbf{r} \right|^2 \right\rangle_\Theta , \qquad (5.7)$$

where the angled brackets represent an orientational average, of the Fourier transform of $\widehat{\beta}_c(\mathbf{r},\mathbf{\Theta})$, and N is proportional to the concentration of the solution.

The formula of eqn (5.7) simplifies further when the solute particles are spherically symmetric; that is to say,

$$\widehat{\beta}_c(\mathbf{r},\mathbf{\Theta}) = \widehat{\beta}_c(r)$$

where $r = |\mathbf{r}|$. Not only does the orientational average then become redundant, but the Fourier transform reduces to a one-dimensional scalar integral:

$$\left(\frac{d\sigma}{d\Omega}\right)_{el} \propto N(4\pi)^2 \left| \int_0^R r^2\, \widehat{\beta}_c(r)\, \frac{\sin(Qr)}{Qr}\, dr \right|^2 , \qquad (5.8)$$

where R is the radius of the object of interest. For the case of eqn (5.5), when $\widehat{\beta}_c(r) = \beta_1 - \beta_0$,

$$\left(\frac{d\sigma}{d\Omega}\right)_{el} \propto N \left[\frac{4\pi\,|\beta_1 - \beta_0|}{Q}\right]^2 \left| \int_0^R r\sin(Qr)\, dr \right|^2 . \qquad (5.9)$$

The integral, which can be done 'by parts', is related to a j_1 spherical Bessel function, plotted in Fig. 5.4,

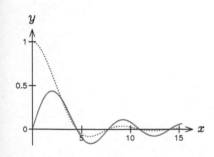

$$\int_0^R r\sin(Qr)\, dr = \frac{\sin(QR) - QR\cos(QR)}{Q^2} = R^2\, j_1(QR) , \qquad (5.10)$$

Fig. 5.4 The spherical Bessel function $y = j_1(x)$ and, dotted, $y = 3\,j_1(x)/x$.

so that eqn (5.9) yields

$$\left(\frac{d\sigma}{d\Omega}\right)_{el} \propto N \left[V_p\, |\beta_1 - \beta_0|\, \frac{3\,j_1(QR)}{QR} \right]^2 , \qquad (5.11)$$

where $V_p = 4\pi R^3/3$ is the volume of the solute particle.

Fourier transforms in polar coordinates

The evaluation of a Fourier transform, or any multiple integral, becomes easier when conducted in a coordinate system that matches the symmetry of the problem. For the three-dimensional case, defined formally in Cartesians in eqns (2.64) and (2.65), it makes most sense to work in spherical polars when $f(\mathbf{r}) = f(r)$. Denoting the angle between the vectors $\mathbf{k}$ and $\mathbf{r}$ by ψ, so that $\mathbf{k} \cdot \mathbf{r} = k r \cos \psi$ with $0 \leqslant \psi \leqslant \pi$ radians, and that of the projection of $\mathbf{r}$ in the plane perpendicular to $\mathbf{k}$ by $0 \leqslant \chi \leqslant 2\pi$, the Fourier transform of $f(r)$ becomes

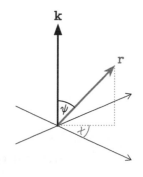

$$F(k) \propto \int_0^R r^2 f(r) \, dr \int_0^\pi \sin \psi \; e^{i k r \cos \psi} \, d\psi \int_0^{2\pi} d\chi \,,$$

with $M = 3$ in eqn (2.65), where the volume element $d^3\mathbf{r} = r^2 \sin \psi \, dr \, d\psi \, d\chi$ and $f(r) = 0$ for $r > R$. Carrying out the χ integral first, which gives 2π, and then the one for ψ,

$$F(k) \propto 2\pi \int_0^R r^2 f(r) \left[\frac{e^{i k r \cos \psi}}{-i k r} \right]_0^\pi dr \,,$$

and recognizing that $e^{i k r} - e^{-i k r} = 2i \sin(kr)$, from eqn (2.32), upon substitution of the ψ limits, yields

$$\iiint f(r) \, e^{i \mathbf{k} \cdot \mathbf{r}} \, d^3\mathbf{r} \;=\; 4\pi \int_0^R r^2 f(r) \, \frac{\sin(kr)}{kr} \, dr \,. \tag{5.12}$$

In the two-dimensional circular case, which is also relevant for cylindrical symmetry, only the angle ψ is needed. While $\mathbf{k} \cdot \mathbf{r}$ is still $k r \cos \psi$, with $0 \leqslant \psi \leqslant 2\pi$ to cover all directions, the element of area $d^2\mathbf{r} = r \, dr \, d\psi$. The corresponding ψ integral is not so straightforward without the prefactor of $\sin \psi$ to accompany the $\cos \psi$ in the exponential but, to within 2π, is a J_0 Bessel function:

$$\int_0^{2\pi} e^{i k r \cos \psi} \, d\psi \;=\; 2\pi J_0(kr) \,. \tag{5.13}$$

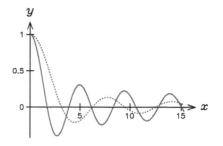

Fig. 5.5 The $y = J_0(x)$ Bessel function and, dotted, the $\sin x / x$ sinc function.

$J_0(x)$ is similar to a $\sin x / x$ sinc function, with the latter also being known as a $j_0(x)$ spherical Bessel function; both are plotted in Fig. 5.5. The two-dimensional form of a Fourier transform in plane polar coordinates then follows readily,

$$\iint f(r) \, e^{i \mathbf{k} \cdot \mathbf{r}} \, d^2\mathbf{r} \;=\; 2\pi \int_0^R r \, f(r) \, J_0(kr) \, dr \,, \tag{5.14}$$

and is illustrated in Fig. 2.19 for the case of $f(r) = 1$ for $r \leqslant R$, and zero otherwise.

5.2.1 Absolute intensities

The intensity of the measured elastic signal, $I_{el}(\mathbf{Q})$, is related to the differential cross-section by

$$I_{el}(\mathbf{Q}) = I_o(\lambda)\, \eta(\lambda)\, T(\lambda)\, \Delta\Omega \left(\frac{d\sigma}{d\Omega}\right)_{el} \otimes R(\mathbf{Q}) + B(\mathbf{Q})\,, \qquad (5.15)$$

where I_o, η and T, which may depend on wavelength, denote the incident flux, efficiency of the detectors and the transmission function of the sample, respectively; $\Delta\Omega$ is the solid angle subtended by the detector-element, corresponding to a wavevector transfer of $\mathbf{Q}$, at the sample position. $B(\mathbf{Q})$ represents a slowly varying background and $R(\mathbf{Q})$ the instrumental resolution function. Strictly speaking the blurring effect of the latter can only be written as a convolution ($\otimes$) for an invariant response, whereas the width of $R(\mathbf{Q})$ generally increases with $|\mathbf{Q}|$, but we'll ignore this technicality for simplicity of notation.

$$\iiint f(\mathbf{q})\, R(\mathbf{q},\mathbf{Q})\, d^3\mathbf{q}$$
$$= f(\mathbf{Q}) \otimes R(\mathbf{Q})$$
$$\text{if } R(\mathbf{q},\mathbf{Q}) = R(\mathbf{Q}-\mathbf{q})$$

Although $T(\lambda)$ will depend on the overall composition of the sample, as might some aspects of $B(\mathbf{Q})$, the details of the structure are encoded in the differential cross-section. At the origin, eqn (5.7) predicts that

$$\left.\left(\frac{d\sigma}{d\Omega}\right)_{el}\right|_{\mathbf{Q}=0} \propto N \left\langle \left| \iiint_{V_\Theta} \widehat{\beta}_c(\mathbf{r},\Theta)\, d^3\mathbf{r} \right|^2 \right\rangle_\Theta. \qquad (5.16)$$

Taking $\widehat{\beta}_c(\mathbf{r},\Theta)$ to be uniform within the volume defined by V_Θ, and equal to $\beta_1-\beta_o$, so that

$$\iiint_{V_\Theta} \widehat{\beta}_c(\mathbf{r},\Theta)\, d^3\mathbf{r} = (\beta_1-\beta_o) \iiint_{V_\Theta} d^3\mathbf{r}\,,$$

and recognizing the integral on the right-hand side as the volume of the solute particle, V_p, irrespective of the orientation Θ,

$$\left.\left(\frac{d\sigma}{d\Omega}\right)_{el}\right|_{\mathbf{Q}=0} \propto N\, V_p^2\, |\beta_1-\beta_o|^2. \qquad (5.17)$$

The SLDs β_o and β_1 are given by eqn (4.11), with a summation over the atoms of the solvent and solute respectively.

The equations for the differential cross-section were written with a proportionality to allow for the variety of normalizations in use. The most common in small-angle scattering studies is 'per unit scattering volume', V, whence

$$\left.\left(\frac{d\sigma}{d\Omega}\right)_{el}\right|_{\mathbf{Q}=0} = \left(\frac{N}{V}\right) V_p^2\, |\beta_1-\beta_o|^2. \qquad (5.18)$$

If the concentration of the solution is defined dimensionlessly by the *volume fraction*, Φ, then the number of solute particles is

$$N = \frac{\Phi V}{V_{\mathrm{p}}} .$$

Their individual volumes, V_{p}, can be related to macroscopic quantities, such as the density of the solute, ρ, and its molecular mass, m, through the Avogadro constant, N_{A}:

$$V_{\mathrm{p}} = \frac{m}{\rho N_{\mathrm{A}}} ,$$

where the units of V_{p} are cm^3 when ρ is given in g cm^{-3} and m in g mol^{-1}. If β_{o} and β_1 are specified in cm^{-2} (or 10^{-16}Å^{-2}), then the differential cross-section,

$$\boxed{\left(\frac{\mathrm{d}\sigma}{\mathrm{d}\Omega}\right)_{\mathrm{el}}\Bigg|_{\mathbf{Q}=0} = \frac{\Phi m}{\rho N_{\mathrm{A}}}\left|\beta_1 - \beta_{\mathrm{o}}\right|^2 ,} \qquad (5.19)$$

will be in units of cm^{-1}.

5.2.2 The Guinier approximation

The relationship between the value of the differential cross-section at $\mathbf{Q}=0$ and the density, molecular mass, concentration and SLD (relative to β_{o}) of the solute, quantified by eqn (5.19), provides a useful constraint on structural models for the sample. The problem, however, is that the intensity of the scattered signal at the origin, $I_{\mathrm{el}}(0)$, is not measured since it coincides with the 'hole' at the centre of the detector (Fig. 5.2). Even without this practical shortcoming, there is a fundamental difficulty with eqn (5.19): it does not hold at $\mathbf{Q}=0$ because it is based on eqn (5.6) in which the $\delta(\mathbf{Q})$ component was omitted! It does apply asymptotically as $\mathbf{Q}\to0$, and an analysis of its behaviour as the origin is approached sheds light on the size of the solute particles.

The first step in the derivation of eqn (5.19) involved the substitution of unity for the exponential term in eqn (5.7). If $\mathbf{Q}\neq0$, but is small, then a Taylor series expansion about the origin gives

$$e^{\mathrm{i}\mathbf{Q}\bullet\mathbf{r}} = 1 + \mathrm{i}\mathbf{Q}\bullet\mathbf{r} + \frac{1}{2!}(\mathrm{i}\mathbf{Q}\bullet\mathbf{r})^2 + \cdots ,$$

where higher-order terms become increasingly negligible as $|\mathbf{Q}|\to0$. Thus eqn (5.7) can be expressed as

$$\left(\frac{\mathrm{d}\sigma}{\mathrm{d}\Omega}\right)_{\mathrm{el}} = \frac{N}{V}\left\langle\left|\sum_{n=0}^{\infty}\iiint_{V_{\Theta}}\widehat{\beta}_{\mathrm{c}}(\mathbf{r},\boldsymbol{\Theta})\frac{(\mathrm{i}\mathbf{Q}\bullet\mathbf{r})^n}{n!}\,\mathrm{d}^3\mathbf{r}\right|^2\right\rangle_{\boldsymbol{\Theta}} , \qquad (5.20)$$

with eqn (5.16) being the case when only the zeroth-order term contributes to the summation. In the limit of small Q, we might expect the linear term to give rise to the main deviation from eqn (5.19). This is not so, however, as

$$\iiint\limits_{V_{\ominus}} (\mathbf{Q}\cdot\mathbf{r})\,\widehat{\beta}_{\mathrm{c}}(\mathbf{r},\mathbf{\Theta})\,\mathrm{d}^3\mathbf{r} = 0$$

$$\mathbf{Q}\cdot(-\mathbf{r}) = -\mathbf{Q}\cdot\mathbf{r}$$

if $\widehat{\beta}_{\mathrm{c}}(\mathbf{r},\mathbf{\Theta}) = \widehat{\beta}_{\mathrm{c}}(-\mathbf{r},\mathbf{\Theta})$. For solute particles with centric symmetry, therefore, the quadratic factor,

$$\mathrm{i}^2 = -1$$

$$-\frac{1}{2}\iiint\limits_{V_{\ominus}} (\mathbf{Q}\cdot\mathbf{r})^2\,\widehat{\beta}_{\mathrm{c}}(\mathbf{r},\mathbf{\Theta})\,\mathrm{d}^3\mathbf{r}\,,$$

dominates the drop-off in scattered intensity away from the origin. Taking $\widehat{\beta}_{\mathrm{c}}(\mathbf{r},\mathbf{\Theta}) = \beta_1 - \beta_{\mathrm{o}}$ within $V_{\ominus}$, as earlier, eqn (5.20) is approximated well by

$$\left(\frac{\mathrm{d}\sigma}{\mathrm{d}\Omega}\right)_{\mathrm{el}} \approx \left(\frac{N}{V}\right) V_{\mathrm{p}}^2\,|\beta_1 - \beta_{\mathrm{o}}|^2 \left\langle \left| 1 - \frac{1}{2V_{\mathrm{p}}}\iiint\limits_{V_{\ominus}} (\mathbf{Q}\cdot\mathbf{r})^2\,\mathrm{d}^3\mathbf{r} \right|^2 \right\rangle_{\ominus}$$

in the neighbourhood of $\mathbf{Q}=0$. Expanding the modulus-squared and using the linearity of the averaging,

$$\left(\frac{\mathrm{d}\sigma}{\mathrm{d}\Omega}\right)_{\mathrm{el}} \approx \left.\left(\frac{\mathrm{d}\sigma}{\mathrm{d}\Omega}\right)_{\mathrm{el}}\right|_{\mathbf{Q}=0} \left[1 - \frac{1}{V_{\mathrm{p}}}\left\langle \iiint\limits_{V_{\ominus}} (\mathbf{Q}\cdot\mathbf{r})^2\,\mathrm{d}^3\mathbf{r} \right\rangle_{\ominus} \right] \qquad (5.21)$$

where only the terms with the lowest order Q-variation have been retained. Denoting the angle between the vectors $\mathbf{Q}$ and $\mathbf{r}$ by ψ, so that $\mathbf{Q}\cdot\mathbf{r} = Q\,r\cos\psi$, the value of ψ (for a given $\mathbf{Q}$ and a fixed point in the solute particle) will vary between 0 and π radians as the orientation $\mathbf{\Theta}$ changes; the angle of the projection of $\mathbf{r}$ in the plane perpendicular to $\mathbf{Q}$, χ say, will correspondingly vary between 0 and 2π. Hence, the orientational average, over a solid angle of 4π steradians, is given by

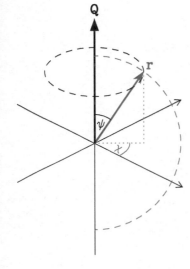

$$\left\langle \iiint\limits_{V_{\ominus}} (\mathbf{Q}\cdot\mathbf{r})^2\,\mathrm{d}^3\mathbf{r} \right\rangle_{\ominus} = \frac{1}{4\pi}\int\limits_0^{2\pi}\mathrm{d}\chi\int\limits_0^{\pi}\sin\psi\,\mathrm{d}\psi\iiint\limits_{V_{\ominus}} (Q\,r\cos\psi)^2\,\mathrm{d}^3\mathbf{r}\,,$$

where $\sin\psi\,\mathrm{d}\psi\,\mathrm{d}\chi$ is the element of solid angle generated by infinitesimally small changes in ψ and χ. Reversing the order of the integrations,

$$\left[\frac{\cos^3\psi}{-3}\right]_0^{\pi} [\chi]_0^{2\pi} = \frac{4\pi}{3}$$

$$\frac{Q^2}{4\pi}\iiint\limits_{V_{\mathrm{p}}} r^2\,\mathrm{d}^3\mathbf{r}\int\limits_0^{\pi}\cos^2\psi\,\sin\psi\,\mathrm{d}\psi\int\limits_0^{2\pi}\mathrm{d}\chi = \frac{Q^2}{3}\iiint\limits_{V_{\mathrm{p}}} r^2\,\mathrm{d}^3\mathbf{r}$$

where the integral of r^2 over the volume of the solute particle is independent of orientation Θ. By analogy with classical mechanics, the latter is related to a *radius of gyration*, R_g, through

$$R_g^2 = \frac{1}{V_p} \iiint_{V_p} r^2 \, d^3 \mathbf{r}, \qquad (5.22)$$

so that

$$\frac{1}{V_p} \left\langle \iiint_{V_\Theta} (\mathbf{Q} \cdot \mathbf{r})^2 \, d^3 \mathbf{r} \right\rangle_{\Theta} = \frac{Q^2 R_g^2}{3}.$$

In conjunction with eqns (5.19) and (5.21), therefore, the logarithm of the differential cross-section in the neighbourhood of the origin is given by

$$\log(XY) = \log X + \log Y$$

$$\boxed{\log_e\left[\left(\frac{d\sigma}{d\Omega}\right)_{el}\right] \approx \log_e\left[\frac{\Phi m}{\rho N_A} \left|\beta_1 - \beta_0\right|^2\right] - \left(\frac{R_g^2}{3}\right)Q^2}, \qquad (5.23)$$

where we have used the first-order term in the Taylor series expansion of $\log_e(1-X)$ to obtain the second contribution on the right-hand side. Thus a plot of $\log_e[d\sigma/d\Omega]$ versus Q^2 in the range $0 < Q \lesssim R_g^{-1}$ should be approximated well by a straight line with a negative slope of $R_g^2/3$ and an intercept related to the molecular mass, concentration and density of the sample.[†]

$$\log_e(1-X) = -X - \frac{X^2}{2} - \cdots$$

[†] Guinier (1939)

The radius of gyration in classical mechanics

The resistance offered by an object to a change in its linear motion, called *inertia*, is determined entirely by its mass, M; the size and shape of the entity have no bearing on the issue (*in vacuo*). The latter become important for rotational motion, however, where the difficulty experienced in trying to change the angular speed is determined by the mass distribution of the object with respect to the axis of rotation. This is encapsulated in the *moment of inertia*,

$$I = \iiint_V r_\perp^2 \, \rho(\mathbf{r}) \, d^3 \mathbf{r},$$

where ρ is the density within volume V and $r_\perp$ is the perpendicular distance of position $\mathbf{r}$ from the axis of rotation. If the density is uniform, so that $\rho(\mathbf{r}) = M/V$, then $I = M R_g^2$ where

$$R_g^2 = \frac{1}{V} \iiint_V r_\perp^2 \, d^3 \mathbf{r}.$$

The radius of gyration, R_g, is then the distance that a point mass M would have to be from the axis of rotation to have the same moment of inertia as the given (extended) object.

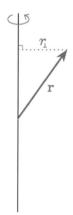

Although the R_g of eqn (5.22) is called a 'radius of gyration', the expression is not quite the same as the one in classical mechanics. This technicality aside, it does give a measure of size in terms of an rms radius. The relationship between R_g and the physical dimensions of the solute particles is simplest to work out for a sphere of radius R:

$$R_g^2 = \frac{1}{(4/3)\pi R^3} \int_0^R r^4 \, \mathrm{d}r \int_0^\pi \sin\theta \, \mathrm{d}\theta \int_0^{2\pi} \mathrm{d}\phi ,$$

where we have used the fact that $\mathrm{d}^3\mathbf{r} = r^2 \sin\theta \, \mathrm{d}r \, \mathrm{d}\theta \, \mathrm{d}\phi$ in spherical polar coordinates, so that

$$\left[\frac{r^5}{5} \right]_0^R \left[-\cos\theta \right]_0^\pi \left[\phi \right]_0^{2\pi} = \frac{4\pi R^5}{5} \qquad\qquad R_g = \sqrt{\frac{3}{5}} R . \qquad (5.24)$$

The resultant fall-off in the the differential cross-section given by eqns (5.23) and (5.24) can be verified by examining the behaviour of eqn (5.11) as $Q \to 0$; that is, by putting

$$\sin x \approx x - \frac{x^3}{6} + \frac{x^5}{120} \quad \text{and} \quad \cos x \approx 1 - \frac{x^2}{2} + \frac{x^4}{24}$$

in eqn (5.10), where $x = QR$, and so on.

For a rugby ball-shaped ellipsoid, with principal axes of half-length a, a and b, it can be shown that

$$R_g = \sqrt{\frac{2a^2 + b^2}{5}} . \qquad (5.25)$$

As required, it reduces to the spherical case of eqn (5.24) when $a = b = R$. A long cylindrical particle can be modelled with $b \gg a$, giving $R_g \approx b/\sqrt{5}$, whereas $a \gg b$ represents a disc-like object, having an $R_g \approx a\sqrt{2/5}$. Since the radius of gyration is dominated by the largest length scale, the Guinier region will be experimentally inaccessible if the solute particles are too extended in some sense, so that $R_g > Q_{min}^{-1}$.

5.2.3   Porod's law

The Guinier relationship of eqn (5.23) quantifies the fact that the scattered signal is insensitive to the shape of the solute particles at the smallest Qs; only their overall size, in an rms sense, and the content of the sample matters. As with all Fourier data, information about structural detail is found at higher Qs. Even then, distinction can be difficult since differences are blurred out by the orientational averaging in eqn (5.7).

The Q-dependence of the differential cross-section for a dilute solution of identical uniform spheres was given in eqn (5.11). The corresponding analytical derivation for particles of other shapes is not so easy, and cannot usually be stated in a closed form; formulae

Table 5.1 Form factors for randomly oriented particles, of uniform scattering-length density contrast $\beta_1 - \beta_0$ and volume $V_{\rm p}$.

Particle shape	Differential cross-section $(\mathrm{d}\sigma/\mathrm{d}\Omega)_{\rm el} / (V_{\rm p}^2 \, |\beta_1 - \beta_0|^2)$
Sphere of radius R	$\left[\dfrac{3 j_1(QR)}{QR} \right]^2$
Thin rod of length L	$\dfrac{2}{QL} \displaystyle\int_0^{QL} \dfrac{\sin x}{x}\, \mathrm{d}x - \dfrac{4 \sin^2(QL)}{(QL)^2}$
Flat disc of radius R	$\dfrac{2}{(QR)^2} \left[1 - \dfrac{J_1(2QR)}{QR} \right]$

for a few idealized cases are given in Table 5.1. When $Q \gg R^{-1}$, the differential cross-section of eqn (5.11) falls off as Q^{-4}:

$$\left(\frac{\mathrm{d}\sigma}{\mathrm{d}\Omega} \right)_{\rm el} \approx \left(\frac{N}{V} \right) \frac{16 \pi^2 R^2}{Q^4} \, |\beta_1 - \beta_0|^2 \cos^2(QR) . \qquad (5.26)$$

By comparison, the scattered intensity at high Qs decays as Q^{-2} and Q^{-1}, respectively, for extremely thin discs and very long rods. The Q^{-4} behaviour noted above for spheres is, in fact, representative of most objects (due to orientational averaging), as long as they are not vanishingly thin in any direction and have a smooth interface with a sharp SLD transition. This result was established theoretically by Porod (1951, 1952), and now bears his name. Averaging over the cosine-squared fringes in eqn (5.26), which gives $1/2$, to mimic the effect of instrumental resolution at high Q, *Porod's law* can be written as

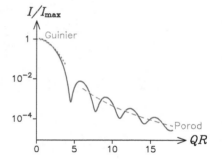

$$\boxed{\left(\frac{\mathrm{d}\sigma}{\mathrm{d}\Omega} \right)_{\rm el} \longrightarrow \left(\frac{N}{V} \right) \frac{2\pi S_{\rm p}}{Q^4} \, |\beta_1 - \beta_0|^2} , \qquad (5.27)$$

for $Q \to \infty$, where $S_{\rm p}$ is the surface area of the particle ($4\pi R^2$ for a sphere).

If the interface between the solute and solvent or, more generally, the regions of the sample with SLDs of β_1 and β_0 is diffuse, then the scattered intensity falls off much more rapidly,

$$\left(\frac{\mathrm{d}\sigma}{\mathrm{d}\Omega} \right)_{\rm el} \sim \frac{\exp(-\sigma^2 Q^2)}{Q^4} ,$$

where σ is an rms measure of the boundary roughness. This is simply a Debye–Waller effect, as discussed earlier in Sections 3.3.2 and 4.1.1.

5.2.4 **Polydispersity**

So far, we have assumed that all the solute particles are identical in size and shape; this is known as *monodispersity*. If there is variation, or *polydispersity*, but no correlation in the locations and orientations, the resultant differential cross-section is just a weighted average of the scattering from the individual constituents. Taking the case of spheres, for example, with a distribution of sizes defined by $f(r)$, so that the fraction with radii between R_1 and R_2 is

$$\int_{R_1}^{R_2} f(r)\,\mathrm{d}r\,,$$

$$\int_0^\infty f(r)\,\mathrm{d}r = 1$$

the differential cross-section is given by

$$\left(\frac{\mathrm{d}\sigma}{\mathrm{d}\Omega}\right)_{\mathrm{el}} = \left(\frac{N}{V}\right)\left[\frac{4\pi\,|\,\beta_1-\beta_0\,|}{Q}\right]^2 \int_0^\infty f(r)\left[r^2 j_1(Qr)\right]^2 \mathrm{d}r\,. \qquad (5.28)$$

Equation (5.11) is recovered with the substitution of $f(r) = \delta(r-R)$, when all the particles have the same radius R.

Polydispersity is usually investigated with a simple model for $f(r)$. A fairly common choice is

$$f(r) = A\,r^\alpha\,\mathrm{e}^{-\gamma r}, \qquad (5.29)$$

which is known as a *Schulz* distribution. Its mean and variance are related to α and γ through

$$\langle r \rangle = \frac{1+\alpha}{\gamma} \quad \text{and} \quad \left\langle (r-\langle r \rangle)^2 \right\rangle = \frac{1+\alpha}{\gamma^2}\,,$$

with the normalization constant being given by $A = \gamma^{1+\alpha}/\Gamma(1+\alpha)$; three examples are shown in Fig. 5.6. Although the integral of eqn (5.28) can be evaluated analytically for a Schulz distribution, the

$$\Gamma(x) = \int_0^\infty t^{x-1}\,\mathrm{e}^{-t}\,\mathrm{d}t$$

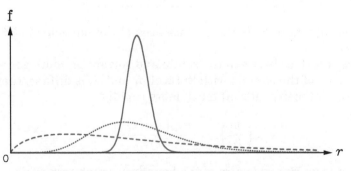

Fig. 5.6 The Schulz distribution: all three examples have the same mean, μ, but standard deviations of $\mu/12$ (solid), $\mu/3$ (dotted) and $3\mu/4$ (dashed).

result is quite messy. In the vicinity of $Q = 0$, however, the formula reduces to a quadratic form reminiscent of eqn (5.23), with R_g now involving the ratio of the sixth and eighth moments of $f(r)$,

$$R_g^2 = \frac{3}{5} \frac{\langle r^8 \rangle}{\langle r^6 \rangle} , \qquad (5.30)$$

and the asymptotic value of the differential cross-section at the origin being given by

$$\left. \left(\frac{\mathrm{d}\sigma}{\mathrm{d}\Omega} \right)_{\mathrm{el}} \right|_{\mathbf{Q}=0} = \left(\frac{N}{V} \right) \left[\frac{4\pi}{3} \left| \beta_1 - \beta_\mathrm{o} \right| \right]^2 \langle r^6 \rangle . \qquad (5.31)$$

As the spread of radial sizes decreases, so that $f(r) \to \delta(r - R)$ and $\langle r^n \rangle \to R^n$, the formulae of eqns (5.30) and (5.31) revert to those of (5.24) and (5.18) respectively. When the polydispersity is substantial, the radius of gyration (and V_p) is weighted towards the larger particles.

5.3   Assemblies and correlations

The discussion of Section 5.2 was based on scattering from a dilute solution of non-interacting particles, or an equivalent system of SLD contrast variation. As the concentration rises, or the solute becomes more polarized, the assumptions of randomly located and oriented scattering centres underlying eqn (5.7) break down.

5.3.1   Orientational alignment

The simplest case of correlation is that of orientational alignment. If liquid samples are subjected to a shear stress (Hayter and Penfold, 1984), or a magnetic or electric field is applied in suitable circumstances, solute particles with non-spherical symmetry can be made to point in the same direction. The formula for the differential cross-section then differs from eqn (5.7) in its lack of averaging over Θ:

$$\left(\frac{\mathrm{d}\sigma}{\mathrm{d}\Omega} \right)_{\mathrm{el}} \propto N \left| \iiint_{V_\Theta} \widehat{\beta}_\mathrm{c}(\mathbf{r}, \Theta) \, \mathrm{e}^{\mathrm{i}\mathbf{Q} \cdot \mathbf{r}} \, \mathrm{d}^3\mathbf{r} \right|^2 . \qquad (5.32)$$

An idealized example of the scattered intensity from identical elliptical particles, of uniform SLD, with principal axes in the ratio 2:1:1 aligned with the x, y and z axes respectively, is shown in Fig. 5.7(a); the incident beam is along z, so that $Q_z \approx 0$, and the central portion of the detector is masked out. The corresponding randomly oriented situation is illustrated in Fig. 5.7(b). The advantage of alignment is clear to see: the non-spherical character of the solute particles is immediately obvious from the anisotropy in Fig. 5.7(a) but is lost in the

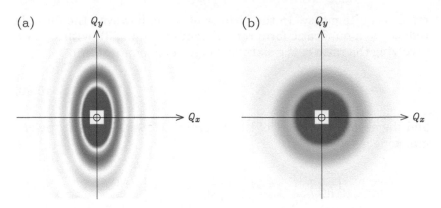

Fig. 5.7 The scattering intensity from a very dilute solution of identical ellipsoids, of uniform SLD, with principal axes in the ratio 2:1:1. The long axis is aligned with the x-direction in (a) and oriented randomly in (b), with the incident beam parallel to z.

integration over Θ. Indeed, the relative spread of the fringes parallel to Q_x and Q_y, and the reciprocal relationship between length scales in real and Fourier space, tells us that the ratio of the long to short axis is 2:1 with the latter being along y. By contrast, the scattering pattern of Fig. 5.7(b) could easily be from a solution of polydispersed spheres.

5.3.2 Positional correlation

Whereas orientational alignment is advantageous for determining the shape of the solute particles, positional correlations make the analysis more difficult. To see this, let's assume orientational (and size) invariance, so that $\widehat{\beta}_{c}(\mathbf{r}, \Theta) = \widehat{\beta}_{c}(\mathbf{r})$, and rewrite the SLD function of eqn (5.3) in the form

$$\beta(\mathbf{r}) = \beta_{o} + \widehat{\beta}_{c}(\mathbf{r}) \otimes \sum_{j=1}^{N} \delta(\mathbf{r}-\mathbf{r}_{j}) . \tag{5.33}$$

With the aid of the convolution theorem of eqn (2.51), the Fourier transform of this $\beta(\mathbf{r})$, and eqn (5.1), gives

$$\left(\frac{d\sigma}{d\Omega}\right)_{el} \propto \left| (2\pi)^3 \delta(\mathbf{Q}) + F(\mathbf{Q}) \sum_{j=1}^{N} e^{i\mathbf{Q}\bullet\mathbf{r}_j} \right|^2 \tag{5.34}$$

where

$$F(\mathbf{Q}) = \iiint_{V} \widehat{\beta}_{c}(\mathbf{r})\, e^{i\mathbf{Q}\bullet\mathbf{r}}\, d^3\mathbf{r} . \tag{5.35}$$

This is very similar to the expression earlier in the chapter, except that previously F also depended on Θ_j and had to remain inside the summation symbol. Ignoring the $\delta(\mathbf{Q})$ contribution when the modulus-squared is evaluated, because the central region close to the origin is not accessible experimentally, the differential cross-section reduces to

$$\left(\frac{\mathrm{d}\sigma}{\mathrm{d}\Omega}\right)_{\mathrm{el}} \propto |\mathrm{F}(\mathbf{Q})|^2 \left\{ N + 2 \sum_{j > j'} \cos\left[\mathbf{Q}\cdot(\mathbf{r}_j - \mathbf{r}_{j'})\right] \right\}, \qquad (5.36)$$

for $\mathbf{Q} \neq 0$. If the locations of the particles are uncorrelated, then the contribution of the $j \neq j'$ summation will be negligible compared to N due to the incoherent addition of the cosine terms; this was assumed in Section 5.2 whence, bar the orientational averaging, eqn (5.36) reduces to eqn (5.7). If there is a discernible pattern in the positions $\{\mathbf{r}_j\}$, however, the differential cross-section is no longer simply proportional to the modulus-squared of the molecular form factor, $\mathrm{F}(\mathbf{Q})$. In that case,

$$\boxed{\left(\frac{\mathrm{d}\sigma}{\mathrm{d}\Omega}\right)_{\mathrm{el}} \propto |\mathrm{F}(\mathbf{Q})|^2 \, \mathcal{S}(\mathbf{Q})} \, , \qquad (5.37)$$

where $\mathcal{S}(\mathbf{Q})$ corresponds to the term in curly brackets in eqn (5.36) and, since it depends on the positional correlations between the solute particles, is known as a *structure factor*.

In small-angle scattering experiments, where the main interest is in the building blocks of the sample, an $\mathcal{S}(\mathbf{Q})$ other than a constant (N) is a complication that is best mitigated through dilution; sometimes, such as with highly polar macromolecules or porous materials, it is just unavoidable. In the study of *amorphous* materials, which are entirely dependent on local interactions and positional correlations, the structure factor plays a central role. We will leave a discussion of $\mathcal{S}(\mathbf{Q})$ to Chapter 6, therefore, which is concerned with that topic.

5.4   Pair-distribution function

The basic scattering equation, (5.1), gives a highly non-linear relationship between the SLD function, $\beta(\mathbf{r})$, and the differential cross-section for elastic scattering, $(\mathrm{d}\sigma/\mathrm{d}\Omega)_{\mathrm{el}}$, due to the modulus-squared. The difficulties resulting from this loss of Fourier phase information, in terms of ascertaining the sample structure from the scattering data easily and uniquely, were highlighted in Section 2.6. As indicated in Section 2.4.2, however, $(\mathrm{d}\sigma/\mathrm{d}\Omega)_{\mathrm{el}}$ does depend linearly on the ACF of $\beta(\mathbf{r})$:

$$\left(\frac{\mathrm{d}\sigma}{\mathrm{d}\Omega}\right)_{\mathrm{el}} \propto \iiint\limits_{V} \mathrm{ACF}[\beta(\mathbf{r})] \, \mathrm{e}^{\mathrm{i}\mathbf{Q}\cdot\mathbf{r}} \, \mathrm{d}^3\mathbf{r} \, , \qquad (5.38)$$

where

$$\text{ACF}\big[\beta(\mathbf{r})\big] = \iiint_V \beta(\mathbf{r}')^* \, \beta(\mathbf{r}'+\mathbf{r}) \, \mathrm{d}^3\mathbf{r}'. \tag{5.39}$$

Thus, scattering data really tell us about the correlations within the SLD function at different separations $\mathbf{r}$.

For randomly located solute particles, when the $\mathcal{S}(\mathbf{Q})$ in eqn (5.37) is a constant, the only coherent contribution to the ACF comes from the internal structure of the macromolecules. If the latter have no preferred orientation, then the ACF is spherically symmetric:

$$\text{ACF}\big[\beta(\mathbf{r})\big] = \begin{cases} \mathrm{g}(r) & \text{for } r < d_{\max}, \\ 0 & \text{otherwise}, \end{cases} \tag{5.40}$$

where $r = |\mathbf{r}|$ and $d_{\max}$ is the longest dimension within the solute particle (e.g. the major axis of an ellipsoid). Following eqn (5.12), therefore, eqn (5.38) reduces to the one-dimensional integral

$$\left(\frac{\mathrm{d}\sigma}{\mathrm{d}\Omega}\right)_{\text{el}} \propto 4\pi \int_0^{d_{\max}} r^2 \, \mathrm{g}(r) \, \frac{\sin(Qr)}{Qr} \, \mathrm{d}r. \tag{5.41}$$

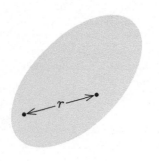

If the particles have a uniform SLD, β_1, then the probability of finding two points within them separated by a distance of r to $r+\mathrm{d}r$ is given by $\mathrm{P}(r)\,\mathrm{d}r$ where

$$\mathrm{P}(r) \propto 4\pi r^2 \mathrm{g}(r). \tag{5.42}$$

The prefactor of $4\pi r^2$ comes from the integration of the ACF of eqn (5.40) over all directions for a given separation r. $\mathrm{P}(r)$ is known as a *pair*, or *distance*, distribution function, and is plotted for several idealized cases in Fig. 5.8.

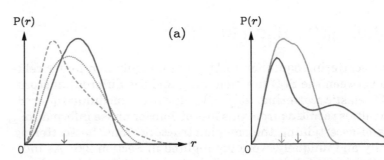

Fig. 5.8 The pair-distribution functions for (a) a sphere (blue) and ellipsoids with principal axes in the ratio of 4:1:1 (dashed) and 1:4:4 (dotted), of uniform SLD, all having the same radius of gyration; (b) composite particles of two identical ellipsoids, with principal axes in the ratio of 2:1:1, joined (long) end-to-end (blue) and side-by-side (grey). The radii of gyration are indicated by arrows on the r-axes.

Mathematically, $P(r)$ can be obtained from the differential cross-section through an inverse Fourier transform:

$$P(r) \propto \frac{2}{\pi} \int_0^\infty Qr \left(\frac{\mathrm{d}\sigma}{\mathrm{d}\Omega}\right)_{\mathrm{el}} \sin(Qr)\, \mathrm{d}Q\,. \tag{5.43}$$

$$\int_0^{d_{\mathrm{max}}} P(r)\, \mathrm{d}r = 1$$

Such a direct inversion of the scattering data is hampered by their incomplete and noisy nature, and the complication that $(\mathrm{d}\sigma/\mathrm{d}\Omega)_{\mathrm{el}}$ is related to the intensity measurements through eqn (5.15) involving a resolution function and background signal; these difficulties were discussed in Section 2.6. Practicalities aside, the radius of gyration of eqn (5.22) can be ascertained from the pair-distribution function through its second moment:

$$R_{\mathrm{g}}^2 = \tfrac{1}{2} \int_0^{d_{\mathrm{max}}} r^2\, P(r)\, \mathrm{d}r\,. \tag{5.44}$$

The factor of a half comes from the fact that the r above refers to the distance between two arbitrary points within the solute particle whereas in eqn (5.22) it denotes the distance of a single point from the centre of symmetry. To see this, let $\mathbf{r}_1$ and $\mathbf{r}_2$ be displacements from the natural origin and $\mathbf{r}_{12}$ be the vector representing the separation between them. Taking the dot product of each side of $\mathbf{r}_{12} = \mathbf{r}_2 - \mathbf{r}_1$ with itself, which gives the cosine formula for a triangle, and averaging over all the different possible pairs of positions within the solute particle,

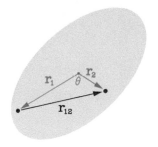

$$\langle r_{12}^2 \rangle = \langle r_1^2 \rangle + \langle r_2^2 \rangle - 2 \langle r_1 r_2 \cos\theta \rangle\,,$$

where θ is the angle between $\mathbf{r}_1$ and $\mathbf{r}_2$. Equation (5.44) follows from the realization that $\langle r_1^2 \rangle = \langle r_2^2 \rangle = R_{\mathrm{g}}^2$, $\langle r_1 r_2 \cos\theta \rangle = 0$ and the left-hand side is the second moment of $P(r)$.

5.5 Contrast matching

For simplicity, we have generally assumed that the macromolecules of interest can be modelled as particles with a uniform SLD, β_1. This is usually adequate since small-angle scattering experiments probe structure on length scales of several atomic diameters, or more, so that only an average SLD is seen. If the object consists of distinct subunits, however, then this approximation may be too poor to explain the scattering data. A common case occurs when a globular molecule is enveloped by a layer of another, so that the structure is better described by a *core-shell* model:

$$\beta_{\mathrm{c}}(\mathbf{r}, \boldsymbol{\Theta}) = \begin{cases} \beta_1 & \text{for } |\mathbf{r}| < R_1\,, \\ \beta_2 & R_1 \leqslant |\mathbf{r}| < R_2\,, \end{cases} \tag{5.45}$$

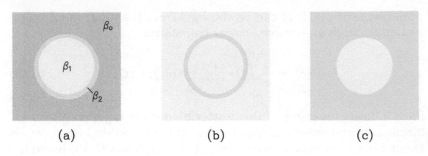

Fig. 5.9 A schematic illustration of contrast matching: (b) $\beta_\mathrm{o} = \beta_1$, and (c) $\beta_\mathrm{o} = \beta_2$.

and zero otherwise. Following eqns (5.4), (5.8) and (5.10), this leads to

$$\left(\frac{\mathrm{d}\sigma}{\mathrm{d}\Omega}\right)_{\mathrm{el}} \propto N \left(\frac{4\pi}{Q}\right)^2 \Big| (\beta_1 - \beta_\mathrm{o})\, R_1^2\, j_1(QR_1) + \tag{5.46}$$
$$(\beta_2 - \beta_\mathrm{o}) \Big[R_2^2\, j_1(QR_2) - R_1^2\, j_1(QR_1) \Big] \Big|^2.$$

The analysis of the data will be simplified, and our ability to distinguish between the two components improved, therefore, if separate measurements can be made with the SLD of the solvent, β_o, chosen to equal β_1 and β_2. Such a *contrast match* is easy to arrange for neutrons, with a suitable mix of hydrogenated and deuterated versions of the solvent. Since H_2O and D_2O have SLDs of $-0.56 \times 10^{-6}\ \text{Å}^{-2}$ and $+6.38 \times 10^{-6}\ \text{Å}^{-2}$, respectively, for example, values of β_o within this range are readily accessible for water-based solvents. The benefits of contrast matching are illustrated schematically for a core-shell structure in Fig. 5.9.

$H \equiv {}^1H$

$D \equiv {}^2H$

Liquids and amorphous materials

6

Liquids and amorphous materials are characterized structurally by their localized order; on length scales larger than a few nanometres, they exhibit no discernible pattern.

6.1 The middle phase of matter

Ice, water and steam are chemically the same material, H_2O (Fig. 6.1), but they differ greatly in their physical properties. Below 0°C, and under normal atmospheric pressure, the water molecules are held in place by a regular and repeating network of *hydrogen bonds*; this solid crystalline ice state is defined by its long-range structural order (Fig. 6.2(a)). At the other extreme, above 100°C, is steam: the random gaseous phase consisting of well-separated molecules. Water occupies the middle ground. While there is enough thermal energy in the system for the hydrogen bonds to be broken easily, there is still an electrochemically driven tendency for neighbouring water molecules to orient themselves in certain directions with re-

Fig. 6.1 With two covalent bonds and two lone pairs per oxygen atom, water molecules are tetrahedral in shape.

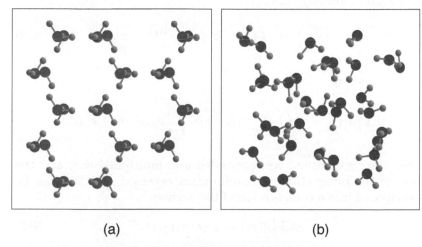

(a)   (b)

Fig. 6.2 (a) Normal ice, with a hexagonal crystalline structure. (b) A snap-shot of water, showing short-range molecular ordering but a generally disordered structure. (Courtesy of the MathMol Project, New York University.)

spect to each other (Fig. 6.2(b)). This leads to some degree of molecular ordering over short length scales, but the correlations become weaker with distance to leave no discernible pattern over separations of more than a few nanometres. Although these structural characteristics are typical of liquids, they can also be found in solids; the latter are known as amorphous materials, with glasses being one of the most common examples.

6.2 Radial distribution functions

Given their lack of a well-defined structure, liquids and amorphous materials are best visualized through their radial distribution functions. We met an example of the latter in Section 5.4, where $P(r)\,dr$ was equal to the probability of finding two points separated by a distance of r to $r+dr$ within the macromolecular object of interest. The main difference with the small-angle scattering case is that the molecules comprising the sample are now small, and densely packed, and the measurements encompass much higher values of Q. As such, $P(r)$ contains information on interatomic distances both within molecules and between them.

To understand the interatomic link better, let us write the SLD function, $\beta(\mathbf{r})$, in the form of eqn (3.34):

$$\beta(\mathbf{r}) = \sum_{j=1}^{N} \beta_j(\mathbf{r}-\mathbf{r}_j)\,, \tag{6.1}$$

where $\beta_j(\mathbf{r})$ is the SLD distribution of atom j when centred at the origin ($\mathbf{r}_j = 0$). The ACF of $\beta(\mathbf{r})$, which is related to the differential cross-section linearly through

$$\left(\frac{d\sigma}{d\Omega}\right)_{\mathrm{el}} \propto \iiint_V \mathrm{ACF}[\beta(\mathbf{r})]\,e^{i\mathbf{Q}\cdot\mathbf{r}}\,d^3\mathbf{r}\,, \tag{6.2}$$

is then given by

$$\mathrm{ACF}[\beta(\mathbf{r})] = \iiint_V \sum_{j=1}^{N} \beta_j(\mathbf{r}'-\mathbf{r}_j)^* \sum_{k=1}^{N} \beta_k(\mathbf{r}'-\mathbf{r}_k+\mathbf{r})\,d^3\mathbf{r}'\,.$$

When the summations are expanded and multiplied out, and the order of the integration and summation reversed, the ACF can be decomposed into a combination of two terms:

$$\mathrm{ACF}[\beta(\mathbf{r})] = g_1(\mathbf{r}) + g_2(\mathbf{r})\,, \tag{6.3}$$

where

$$g_1(\mathbf{r}) = \sum_{j=1}^{N} \iiint_V \beta_j(\mathbf{r}'-\mathbf{r}_j)^* \beta_j(\mathbf{r}'-\mathbf{r}_j+\mathbf{r})\,d^3\mathbf{r}' \tag{6.4}$$

and

$$g_2(\mathbf{r}) = \sum_{j \neq k} \iiint_V \beta_j(\mathbf{r'} - \mathbf{r}_j)^* \, \beta_k(\mathbf{r'} - \mathbf{r}_k + \mathbf{r}) \, \mathrm{d}^3 \mathbf{r'}. \qquad (6.5)$$

The first, $g_1(\mathbf{r})$, is fairly uninteresting, as it results from the correlation of each atom with itself. It has a maximum at the origin,

$$g_1(0) = \sum_{j=1}^{N} \iiint \left| \beta_j(\mathbf{r'}) \right|^2 \mathrm{d}^3 \mathbf{r'},$$

and decays rapidly to zero with increasing r; the fall-off occurs over an atomic diameter for X-rays and a much shorter distance, determined by the rms thermal motions of the atoms, for neutrons. The second term, $g_2(\mathbf{r})$, depends on the spatial distribution of the atoms, and their types, and conveys information on the structure of the sample; let's examine this further.

For a sample consisting of N atoms, eqn (6.5) entails the summation of $N(N-1)$ functions of r. Each of these is the result of an integral that represents the correlation between atom j and atom k, with $j \neq k$, which will have a maximal value (in absolute terms) of

$$\iiint \beta_j(\mathbf{r'})^* \, \beta_k(\mathbf{r'}) \, \mathrm{d}^3 \mathbf{r'}$$

when $\mathbf{r} = \mathbf{r}_k - \mathbf{r}_j$ and then decay to zero (within an atomic diameter for X-rays, or a few times the rms thermal amplitude for neutrons). There will be an equivalent contribution when the suffices j and k are interchanged, because the same two atoms are involved, so that $g_2(\mathbf{r})$ will be symmetric:

$$g_2(-\mathbf{r}) = g_2(\mathbf{r})^*. \qquad (6.6)$$

We have included the complex conjugates in the analysis for strict technical correctness, but they are unnecessary in our implicitly assumed case of real SLDs. While the relationship of eqn (6.6) holds for any type of sample, $g_2(\mathbf{r})$ has to be spherically symmetric for liquids and amorphous materials,

$$g_2(\mathbf{r}) = g_2(r)$$

where $r = |\mathbf{r}|$, as the constituent molecules can be found in all orientations; this symmetry also applies to $g_1(\mathbf{r})$.

The discussion above indicates that $g_2(r)$ will contain a significant signal only at values of r that correspond to interatomic distances present in the sample. This will always be positive for X-rays, but can be negative for neutrons since the scattering lengths of the relevant pair of atoms can have opposite signs. The 'peaks' closest to the origin will be defined most clearly, and will stem from the fixed intra-molecular atomic distances and the nearest-neighbour inter-molecular ones. In water, for example, these would be the covalent

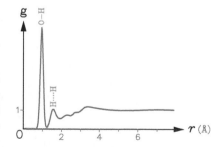

Fig. 6.3 A simulation of the neutron pair-correlation function, $g(r)$, for deuterated water. (Soper, 2000).

O–H and hydrogen OⅢⅢH bond lengths, at $r = 1.0\,\text{Å}$ and $r \approx 1.9\,\text{Å}$ respectively, and the H$\cdots$H distance of $r = 1.6\,\text{Å}$ (Fig. 6.3). As r increases beyond a few nanometres, the lack of a long-range order will lead to a uniform continuum of interatomic distances. We can estimate the asymptotic value of $g_2(r)$ by substituting the average value of the SLD, $\langle \beta \rangle$, into eqn (5.39):

$$g_2(r) \longrightarrow \langle \beta \rangle^2\, V,$$

where V is the 'illuminated' volume of the sample (assumed small compared to the total).

To aid a comparison between different samples, it's helpful to work with a normalized version of $g_2(r)$ which tends to unity for large r:

$$g(r) = \frac{g_2(r)}{\langle \beta \rangle^2\, V}. \tag{6.7}$$

This is called a *pair-correlation*, or a pair-distribution, function but is often simply referred to as $g(r)$. Other functions that are closely related and commonly used include

$$P(r) = 4\pi r^2\, g(r) \quad \text{and} \quad T(r) = 4\pi r\, g(r). \tag{6.8}$$

They all convey the same structural information, and really only differ in their asymptotic behaviour at large r: $P(r) \propto r^2$, $T(r) \propto r$ and $g(r) \to 1$. While only $P(r)$ can strictly be called a radial distribution function, since it entails the integral of a three-dimensional function, $g(\mathbf{r})$, over all directions for a given value of $|\mathbf{r}|$,

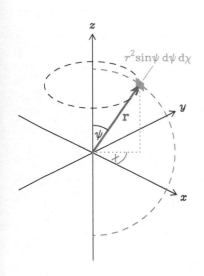

$$P(r) = \int_0^{2\pi} d\chi \int_0^{\pi} g(\mathbf{r})\, r^2 \sin\psi\, d\psi,$$

which reduces to $4\pi r^2\, g(r)$ as $g(\mathbf{r})$ is independent of the spherical polar angles χ and ψ, $T(r)$ and $g(r)$ can also be considered as such because they too depend only on $|\mathbf{r}|$. The three functions, $g(r)$, $T(r)$ and $P(r)$, are illustrated in Fig. 6.4.

6.3 Structure factors

The linear relationship between the differential cross-section and the ACF of the SLD function enshrined in eqn (6.2) means that the decomposition of the latter into $g_1(\mathbf{r})$ and $g_2(\mathbf{r})$ in eqn (6.3) must be mirrored in the scattering data. To see this, let's start by taking the Fourier transform of $\beta(\mathbf{r})$ in eqn (6.1):

$$\iiint_V \beta(\mathbf{r})\, e^{i\mathbf{Q}\bullet\mathbf{r}}\, d^3\mathbf{r} = \sum_{j=1}^{N} f_j(\mathbf{Q})\, e^{i\mathbf{Q}\bullet\mathbf{r}_j}, \tag{6.9}$$

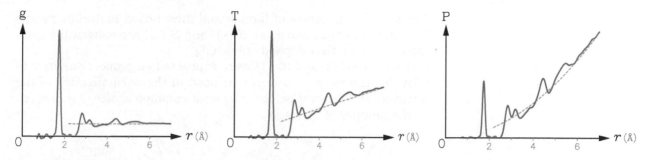

Fig. 6.4 The radial distribution functions $g(r)$, $T(r)$ and $P(r)$, as defined in eqn (6.8), for germania glass (vitreous GeO_2); the expected asymptotic behaviour at large r is indicated by the dashed grey lines. (Hannon *et al.*, 1990)

where atom j is located at $\mathbf{r}_j$ and has a form factor $f_j(\mathbf{Q})$,

$$f_j(\mathbf{Q}) = \iiint \beta_j(\mathbf{r})\, e^{i\mathbf{Q}\cdot\mathbf{r}}\, d^3\mathbf{r}. \qquad (6.10)$$

Form factors were previously discussed in Sections 3.2 and 3.3, and denoted by $f_j(\lambda, \theta)$. Since the differential cross-section is given by the modulus-squared of the Fourier transform of $\beta(\mathbf{r})$, as in eqn (5.1), the product of the sum on the right-hand side of eqn (6.9) with its complex conjugate leads to

$$\left(\frac{d\sigma}{d\Omega}\right)_{el} \propto \mathcal{S}_1(\mathbf{Q}) + \mathcal{S}_2(\mathbf{Q}), \qquad (6.11)$$

where

$$\mathcal{S}_1(\mathbf{Q}) = \sum_{j=1}^{N} \left| f_j(\mathbf{Q}) \right|^2 \qquad (6.12)$$

and

$$\mathcal{S}_2(\mathbf{Q}) = 2 \sum_{j>k} \mathcal{R}e\left\{ f_j(\mathbf{Q})\, f_k(\mathbf{Q})^*\, e^{i\mathbf{Q}\cdot(\mathbf{r}_j - \mathbf{r}_k)} \right\}. \qquad (6.13)$$

The first term depends purely on the scattering properties of individual atoms, and not on their locations; as such, $\mathcal{S}_1(\mathbf{Q})$ is the Fourier counterpart of $g_1(\mathbf{r})$,

$$\mathcal{S}_1(\mathbf{Q}) = \iiint_V g_1(\mathbf{r})\, e^{i\mathbf{Q}\cdot\mathbf{r}}\, d^3\mathbf{r},$$

and is of little intrinsic interest. Structural information about the sample is contained in $\mathcal{S}_2(\mathbf{Q})$, since it depends on the relative positions of different pairs of atoms, which makes it the Fourier counterpart of $g_2(\mathbf{r})$:

$$\mathcal{S}_2(\mathbf{Q}) = \iiint_V g_2(\mathbf{r})\, e^{i\mathbf{Q}\cdot\mathbf{r}}\, d^3\mathbf{r}. \qquad (6.14)$$

The disordered nature of liquids and amorphous materials means that, just like $g_1(\mathbf{r})$ and $g_2(\mathbf{r})$, $\mathcal{S}_1(\mathbf{Q})$ and $\mathcal{S}_2(\mathbf{Q})$ are spherically symmetric; that is, they depend only on $|\mathbf{Q}|$.

Equations (6.2) and (6.11) were expressed as proportionalities to allow for the various conventions used in the normalization of the differential cross-section. For the most common choice of a division by the number of atoms, N,

$$\left(\frac{d\sigma}{d\Omega}\right)_{el} = \frac{1}{N}\sum_{j=1}^{N}\left|f_j(Q)\right|^2 + \frac{4\pi\langle\beta\rangle^2 V}{N}\int_0^R r^2\, g(r)\,\frac{\sin(Qr)}{Qr}\,dr\,,$$

where eqns (5.12) and (6.7) have been used to rewrite the Fourier integral of eqn (6.14) in its one-dimensional form and $R=(3V/4\pi)^{1/3}$. The first term on the right-hand side is simply the average value of the modulus-squared of the form factor, $\langle|f(Q)|^2\rangle$, and the second is better evaluated by calculating the Fourier transform of $g(r)-1$ instead of $g(r)$:

$$\int_0^R r^2\left[g(r)-1\right]\frac{\sin(Qr)}{Qr}\,dr = \int_0^R r^2\, g(r)\,\frac{\sin(Qr)}{Qr}\,dr + \Delta(Q)$$

where the discrepancy $\Delta(Q)$, being the Fourier transform of a constant (unity), is peaked sharply around the origin ($Q \lesssim 2\pi/R$). Since the region close to $Q=0$ is not accessible experimentally, as it coincides with the straight-through unscattered beam, and because the upper limit of the integral on the left-hand side can be replaced with infinity as $g(r)-1 \to 0$ for large r,

$$\left(\frac{d\sigma}{d\Omega}\right)_{el} = \left\langle|f(Q)|^2\right\rangle \mathcal{S}(Q)\,, \tag{6.15}$$

where the structure factor, $\mathcal{S}(Q)$, is given by

$$\boxed{\mathcal{S}(Q) = 1 + \frac{4\pi\alpha(Q)}{Q}\int_0^\infty r\left[g(r)-1\right]\sin(Qr)\,dr}\,, \tag{6.16}$$

for $Q\neq0$, where

$$\alpha(Q) = \frac{\langle\beta\rangle^2 V}{\langle|f(Q)|^2\rangle N}\,. \tag{6.17}$$

The expression for α is independent of Q for neutrons, as $f_j(Q)$ is simply equal to the (coherent) scattering length of the jth nucleus, $\langle b_j\rangle$. With $\langle\beta\rangle = n\langle b\rangle$, where $n=N/V$ is the atomic number density, the formula for α reduces to just

$$\alpha(Q) = n$$

for neutron scattering from a monatomic sample.

Mathematically, eqn (6.16) can be rearranged to give $g(r)$ in terms of $\mathcal{S}(Q)$:

$$g(r) = 1 + \frac{1}{2\pi^2 r} \int_0^\infty Q \, \frac{[\mathcal{S}(Q) - 1]}{\alpha(Q)} \, \sin(Qr) \, \mathrm{d}Q \,.$$

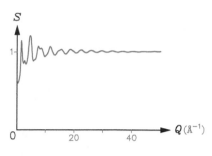

Fig. 6.5 The neutron scattering data related to the pair-correlation functions of Fig. 6.4. (Hannon *et al.*, 1990)

As a means of obtaining the pair-correlation functions from the scattering data, however, such a direct inversion is fraught with many practical difficulties; these were discussed in Section 2.6. While the lack of phase information in diffraction data is not an issue for $g(r)$, the limited extent of the measurements in Q, their noisy nature, the blurring from the instrumental resolution, the (incoherent) background signal, and other experimental considerations, all need to be taken into account.

6.3.1 Partial structure factors

The pair-correlation function is easiest to interpret for a monatomic sample, as all the 'peaks' in $g(r)$ correspond to distances between atoms of the same type. The situation becomes moderately complicated even for the diatomic case, such as water or a binary alloy, since $g(r)$ is then a conglomeration of three sets of separations: AA, AB and BB, where the two elements have generically been labelled as A and B. There are six pairwise combinations for a triatomic sample, and so on.

For a material consisting of M different types of atoms, $g(r)$ can be decomposed into a linear combination of $M(M+1)/2$ partial pair-correlation functions, $g_{JK}(r)$:

$$g(r) = \sum_{J=1}^{M} \sum_{K=J}^{M} \gamma_{JK} \, g_{JK}(r) \,, \tag{6.18}$$

where the coefficients, γ_{JK}, are fixed by the requirements of consistency with eqns (6.5) and (6.7). With $g_{JK}(r) \to 1$ for large r, $g(r)$ can only do likewise if the coefficients sum to unity:

$$\sum_{J=1}^{M} \sum_{K=J}^{M} \gamma_{JK} = 1 \,. \tag{6.19}$$

In conjunction with the observation that

$$\gamma_{JK} \propto (2 - \delta_{JK}) \, c_J \, c_K \, \langle b_J \rangle \, \langle b_K \rangle \,, \tag{6.20}$$

where the *Kronecker* delta, δ_{JK}, equals 1 if $J = K$ and 0 otherwise, and c_J is the fraction of atoms of type J in the sample, the constant of proportionality is found to be the inverse of the square of the average scattering length, $\langle b \rangle^{-2}$.

$$\langle b \rangle = \sum_{J=1}^{M} c_J \, \langle b_J \rangle$$

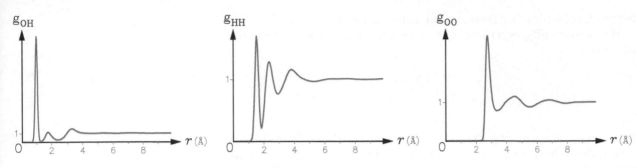

Fig. 6.6 The partial pair-correlation functions, $g_{OH}(r)$, $g_{HH}(r)$ and $g_{OO}(r)$, corresponding to the structural simulation of water in Fig. 6.3. (Soper, 2000)

The decomposition of eqn (6.18) is mirrored in Q-space, with $S(Q)$ being a linear combination of $M(M+1)/2$ partial structure factors, $S_{JK}(Q)$:

$$S(Q) = \sum_{J=1}^{M} \sum_{K=J}^{M} \gamma_{JK}\, S_{JK}(Q)\,. \qquad (6.21)$$

This equation, and the link between the partial structure and pair-correlation functions,

$$S_{JK}(Q) = 1 + \frac{4\pi\,\alpha(Q)}{Q} \int_{0}^{\infty} r\left[g_{JK}(r)-1\right]\sin(Qr)\,\mathrm{d}r\,, \qquad (6.22)$$

is easily established by substituting for $g(r)$ from eqn (6.18), and for unity from eqn (6.19), into eqn (6.16).

The analysis above tells us that information on the distribution of distances between atoms of type J and K, enshrined in $g_{JK}(r)$, is contained unambiguously in the corresponding partial structure factor, $S_{JK}(Q)$. To see how the latter could be ascertained from measurements of the differential cross-section, let's explicitly consider the case of a binary sample. With atoms of only type A and B, eqn (6.21) becomes

$$S(Q) = \frac{c_{A}^{2}\langle b_{A}\rangle^{2}\,S_{AA}(Q) + c_{B}^{2}\langle b_{B}\rangle^{2}\,S_{BB}(Q) + 2\,c_{A}\,c_{B}\langle b_{A}\rangle\langle b_{B}\rangle\,S_{AB}(Q)}{\left[c_{A}\langle b_{A}\rangle + c_{B}\langle b_{B}\rangle\right]^{2}}\,,$$

so that the scattering data, $S(Q)$, are a (known) linear combination of the three partial structure factors $S_{AA}(Q)$, $S_{BB}(Q)$ and $S_{AB}(Q)$. If it were possible to alter the scattering lengths of the atoms, $\langle b_{A}\rangle$ or $\langle b_{B}\rangle$, or both, while keeping everything else the same, then we would obtain scattering data pertaining to a different weighted sum of the desired partial structure factors. In the absence of experimental limitations, three such sets of measurements would be sufficient to determine $S_{AA}(Q)$, $S_{BB}(Q)$ and $S_{AB}(Q)$ uniquely since, for a given Q, it would amount to solving three linear simultaneous equations for three unknowns. A generalization of this argument leads to the conclusion that the partial structure factors of an M-element sample could be disentangled in this way by combining the data from $M(M+1)/2$ suitable differential cross-sections. The nuclear dependence of neutron scattering lengths makes this a viable strategy if sufficient well-calibrated isotopic variations of the sample can be prepared.

6.4 Comparison with small-angle scattering

We began our discussion of liquids and amorphous materials by noting that they represent a state half way between perfect order and complete disorder. Samples studied with small-angle scattering, by contrast, are ideally considered to have a gaseous structure. This distinction lies at the heart of the apparent difference between the associated analyses.

The common theoretical origin of the material in Chapters 5 and 6 is underlined by the similarity of eqns (5.37) and (6.15), where the differential cross-section is expressed as the product of a form factor term and a structure factor. The former is defined by the nature of the basic constituents of the sample, whereas the latter depends on their relative locations. The analysis of small-angle scattering experiments is simplified through the 'dilute solution' approximation whereby the macromolecules, or density fluctuations, of interest are spread so far apart as to be non-interacting. This is, in essence, the ideal gas model. Without positional correlations, the structure factor is featureless and the differential cross-section depends only on the size and shape of the constituent particles. The assumption of a dilute solution becomes poorer as the concentration rises, and eventually the approximation of a uniform structure factor is no longer adequate. Both $\mathcal{S}(Q)$ and the form factor then play an important role in our structural understanding of the sample from small-angle scattering measurements.

In the study of liquids and amorphous materials, the constituent particles are composed of only a few atoms, sometimes just one, and they are not separated by an intervening solvent or air-like matrix. The form factor term is of little interest, therefore, as it simply tells us an average property of the atoms in the sample. For neutrons, it merely constitutes a multiplicative constant as far as the differential cross-section is concerned; for X-rays, it's a modulation that decays slowly with Q. The structural information is contained entirely in $\mathcal{S}(Q)$ which, in turn, is a reflection of the local atomic interactions in the system.

6.5 The Placzek correction

For simplicity, the analysis in this chapter, as in the preceding three, has been based on the assumption of elastic scattering. That is to say, it hinged upon the Fourier relationship between the differential cross-section and the SLD function derived in Chapter 3:

$$\left(\frac{d\sigma}{d\Omega}\right)_{el} \propto \left| \iiint_V \beta(\mathbf{r})\, e^{i\mathbf{Q}\cdot\mathbf{r}}\, d^3\mathbf{r} \right|^2 , \tag{6.23}$$

where the wavevector transfer is given by eqn (3.8). As noted in Section 3.1.3, diffraction experiments generally don't involve energy discrimination and rely on the property that the scattering tends to be predominantly elastic:

$$\frac{d\sigma}{d\Omega} = \left(\frac{d\sigma}{d\Omega}\right)_{el} + \left(\frac{d\sigma}{d\Omega}\right)_{inel} \approx \left(\frac{d\sigma}{d\Omega}\right)_{el} . \tag{6.24}$$

While inelastic events can be ignored to a very good approximation in solid samples, this is not the case for liquids since their atomic structure cannot reasonably be regarded as being fixed. A proper discussion of the issue will have to wait until Chapter 8, when the basic ideas of inelastic scattering are introduced, but we state here that eqn (6.23) holds for the total differential cross-section as long as the energy transferred is small compared to the incident kinetic energy of the X-ray or neutron (so that $|\mathbf{k}_i| \approx |\mathbf{k}_f|$); this is known as the *static approximation*. Under this condition of 'virtual elasticity', $d\sigma/d\Omega$ is related to the Fourier transform of an instantaneous view of the ever changing detailed structure.

Following the work of Placzek (1952), a correction is usually applied to scattering data from liquids to account for deviations from the static approximation. The related multiplicative factor is generally small and becomes increasingly negligible for heavier atoms and higher incident energies.

Periodicity, symmetry and crystallography

7

Our final look at elastic scattering concerns the study of the most highly ordered phase of matter: the crystalline state. This gives rise to diffraction patterns with very sharp and clearly defined structure. For this reason, it is the field in which the earliest (X-ray) scattering experiments were conducted.

7.1 Repetitive structures and Bragg peaks

The central property of crystalline materials, and the feature that endows them with long-range order, is their inherently repetitive nature (Fig. 7.1). This underlying periodicity in the structure can be stated mathematically as

$$\beta(\mathbf{r}) = \beta(\mathbf{r} + n_1\mathbf{a} + n_2\mathbf{b} + n_3\mathbf{c}) \tag{7.1}$$

for any set of integers n_1, n_2 and n_3. That's to say, the SLD at a given location in the sample, $\mathbf{r}$, is the same as that at a point translated by integer multiples of the vectors a, b and c with respect to it. The parallelepiped formed by a, b and c defines the *unit cell*, or basic building block, of the crystal: identical copies are stacked in a three-dimensional lattice to create the periodic structure. The lengths of the vectors (a, b, c) and the angles between them (α, β, γ) are known as *lattice constants*.

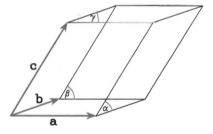

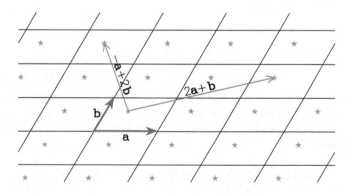

Fig. 7.1 A two-dimensional illustration of the fundamental periodicity in crystals: the stars mark equivalent points, related by translational symmetry.

As always, the principal equation linking the SLD function to the differential cross-section for elastic scattering is

$$\left(\frac{d\sigma}{d\Omega}\right)_{el} \propto \left| \iiint_V \beta(\mathbf{r}) \, e^{i\mathbf{Q}\cdot\mathbf{r}} \, d^3\mathbf{r} \right|^2 , \tag{7.2}$$

where V is the volume of the sample illuminated by the beam of X-rays or neutrons. For a $\beta(\mathbf{r})$ with the translational symmetry in eqn (7.1), the Fourier transform is related to an integral over the volume of a single unit cell, V_{cell}:

$$\iiint_V \beta(\mathbf{r}) \, e^{i\mathbf{Q}\cdot\mathbf{r}} \, d^3\mathbf{r} = L_R(\mathbf{Q}) \iiint_{V_{cell}} \beta(\mathbf{r}) \, e^{i\mathbf{Q}\cdot\mathbf{r}} \, d^3\mathbf{r} \tag{7.3}$$

where

$$L_R(\mathbf{Q}) = \sum_{n_1} \sum_{n_2} \sum_{n_3} \exp\big[i\mathbf{Q}\cdot(n_1\mathbf{a} + n_2\mathbf{b} + n_3\mathbf{c})\big] . \tag{7.4}$$

The summation of the imaginary exponential over n_1, n_2 and n_3 will cancel out to zero unless the terms add up coherently. For the latter to happen, $\mathbf{Q}$ must satisfy the condition

$$e^{i2\pi n} = 1$$

$$\mathbf{Q}\cdot(n_1\mathbf{a} + n_2\mathbf{b} + n_3\mathbf{c}) = \phi_o + 2\pi n , \tag{7.5}$$

where ϕ_o is a constant and n is an integer. This is achieved when

$$\mathbf{Q} = h\mathbf{A} + k\mathbf{B} + l\mathbf{C} \tag{7.6}$$

with h, k and l being integers and

$$\boxed{\mathbf{A} = \frac{2\pi\,\mathbf{b}\times\mathbf{c}}{\mathbf{a}\cdot(\mathbf{b}\times\mathbf{c})}, \quad \mathbf{B} = \frac{2\pi\,\mathbf{c}\times\mathbf{a}}{\mathbf{a}\cdot(\mathbf{b}\times\mathbf{c})}, \quad \mathbf{C} = \frac{2\pi\,\mathbf{a}\times\mathbf{b}}{\mathbf{a}\cdot(\mathbf{b}\times\mathbf{c})}} . \tag{7.7}$$

$$V_{cell} = |\mathbf{a}\cdot(\mathbf{b}\times\mathbf{c})|$$

Using the properties of *scalar triple products*, the vectors $\mathbf{A}$, $\mathbf{B}$ and $\mathbf{C}$ can be shown to obey

$$\begin{aligned} \mathbf{a}\cdot\mathbf{A} = \mathbf{b}\cdot\mathbf{B} = \mathbf{c}\cdot\mathbf{C} = 2\pi \quad \text{and} \\ \mathbf{a}\cdot\mathbf{B} = \mathbf{a}\cdot\mathbf{C} = \mathbf{b}\cdot\mathbf{A} = \mathbf{b}\cdot\mathbf{C} = \mathbf{c}\cdot\mathbf{A} = \mathbf{c}\cdot\mathbf{B} = 0 , \end{aligned} \tag{7.8}$$

so that $n = n_1 h + n_2 k + n_3 l$ and $\phi_o = 0$ in eqn (7.5). Hence, $L_R(\mathbf{Q})$ will have a value V/V_{cell}, the number of illuminated unit cells, if $\mathbf{Q}$ satisfies eqn (7.6) but will be vanishingly small otherwise. As such, the non-zero points of $L_R(\mathbf{Q})$ define a regular three-dimensional grid in $\mathbf{Q}$-space,

$$L_R(\mathbf{Q}) \propto \sum_h \sum_k \sum_l \delta\big(\mathbf{Q} - (h\mathbf{A} + k\mathbf{B} + l\mathbf{C})\big) , \tag{7.9}$$

which is known as the *reciprocal lattice*; the vectors $\mathbf{A}$, $\mathbf{B}$ and $\mathbf{C}$ are called *reciprocal vectors*.

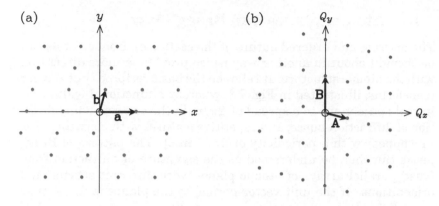

Fig. 7.2 (a) A two-dimensional crystal lattice, L(r), defined by the unit cell vectors a and b. (b) The corresponding reciprocal lattice, $L_R(Q)$, with reciprocal vectors A and B: a is perpendicular to B, b is orthogonal to A and their lengths satisfy the inverse relationship $A/B = b/a$.

The conclusions of the preceding analysis, that the scattering from a crystalline sample is non-zero only at sharp well-defined points in Q, can also be arrived at with the convolution theorem of Section 2.4.1. To see this, we simply need to note that the repetitive structure encapsulated by eqn (7.1) can be expressed as a convolution of a single unit cell and a three-dimensional array of δ-functions defining the crystal lattice, L(r):

$$\beta(\mathbf{r}) = L(\mathbf{r}) \otimes \beta_{cell}(\mathbf{r}) \qquad (7.10)$$

where

$$L(\mathbf{r}) = \sum_{n_1}\sum_{n_2}\sum_{n_3} \delta(\mathbf{r} - (n_1\mathbf{a} + n_2\mathbf{b} + n_3\mathbf{c})). \qquad (7.11)$$

According to the convolution theorem, therefore, the Fourier transform of $\beta(\mathbf{r})$ will be equal to the product of the Fourier transforms of L(r) and the SLD of a unit cell. This is exactly eqn (7.3), with the identification that

$$L_R(\mathbf{Q}) = \iiint_V L(\mathbf{r})\, e^{i\mathbf{Q}\bullet\mathbf{r}}\, d^3\mathbf{r}. \qquad (7.12)$$

While the quantitative result of eqns (7.7) and (7.9) follows from the earlier argument, it can be understood qualitatively as being a generalization of the diffraction pattern from a grating given in Section 2.5.3: it's just a three-dimensional array of δ-functions with 'reciprocal spacing'. Although the Fourier transform of $\beta_{cell}(\mathbf{r})$ is defined for all Q, the discrete nature of $L_R(Q)$ with which it is multiplied means that the scattering from an ideal crystalline sample can be non-zero only for very specific values of Q. These isolated points of scattered intensity are known as *Bragg peaks*.

7.1.1   Atomic planes and Bragg's law

The discrete and ordered nature of the scattering from crystals can be thought about in another way which provides a more direct link with the atomic structure. It relies on the basic reciprocity of Fourier transforms, illustrated in Fig. 7.3, whereby a functional width of w in real-space maps to a spread of $2\pi/w$ in the corresponding direction of diffraction-space, and repetitive features with a spacing of d in r appear with a periodicity of $2\pi/d$ in $\mathbf{Q}$. The pattern of Bragg peaks can then be understood as the signature of diffraction from (vast) parallel arrays of atomic planes, with different spacing and orientations. If the unit vector normal to the planes is $\hat{\mathbf{n}}$ and they are a distance d apart, so that

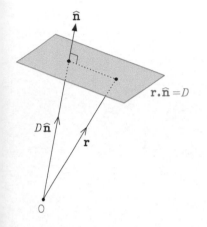

$$\mathbf{r}\boldsymbol{\cdot}\hat{\mathbf{n}} = Nd + \Delta \tag{7.13}$$

where Δ is an offset and $N = 0, \pm1, \pm2, \pm3, \ldots$, they will give rise to a line of diffraction peaks at

$$\mathbf{Q} = \frac{2\pi N}{d}\,\hat{\mathbf{n}}\,. \tag{7.14}$$

When the modulus of this wavevector transfer is equated with the expression in eqn (3.7),

$$Q = \frac{4\pi \sin\theta}{\lambda} = \frac{2\pi N}{d}\,,$$

it leads to the most well-known result in scattering theory:

$$\boxed{N\lambda = 2\,d\,\sin\theta}\,, \tag{7.15}$$

or *Bragg's law*.

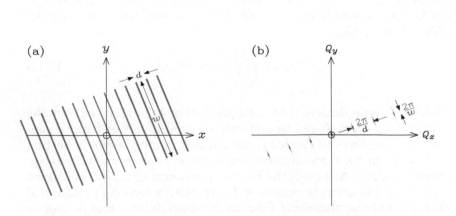

Fig. 7.3 (a) The aperture function for a two-dimensional grating of line spacing d and longitudinal width w. (b) The corresponding diffraction pattern.

An elementary derivation of Bragg's law

The idea that X-rays might be diffracted by a crystal lattice was first proposed by Max von Laue (1912); it was subsequently observed by Friedrich and Knipping (1913), and confirmed the wave nature of X-rays. Lawrence Bragg (1912) extended this work by providing an elementary argument that helped with the understanding of the scattering pattern and, along with his father Henry Bragg (1913), went on to build an X-ray spectrometer and solve some simple crystal structures.

Bragg imagined that, theoretically, the situation would be similar to the case of specular reflection from an array of parallel planes. Following the geometrical setup in Fig. 7.4, with an interplanar spacing of d and a scattering angle of 2θ, the difference in the length of the path travelled by waves reflected from adjacent planes, ΔL, is

$$\Delta L = 2\, d \sin \theta.$$

Constructive interference between all the outgoing waves will only occur, however, when this path difference is a whole number of wavelengths, λ:

$$\Delta L = N\lambda,$$

where N is an integer. The combination of these equations leads to the Bragg condition for a scattering signal to be detected:

$$N\lambda = 2\, d \sin \theta.$$

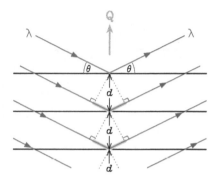

Fig. 7.4 Bragg's setup for reflections from uniformly spaced layers.

Although eqn (7.15) provides a simple relationship between the interplanar spacing and the resultant scattering angles for X-rays, or neutrons, of a given wavelength, it loses the vectorial content of eqn (7.14). Specifically, that the direction of $\mathbf{Q}$ is perpendicular to the atomic planes responsible for the scattered signal. The wavevector transfer can always be defined by two scattering angles (θ and ϕ) and λ, as in Fig. 3.3 and eqn (3.8), but in the crystallographic case it is also specified by the three integers in eqn (7.6). Hence, the Bragg peak indexed by h, k and l (with $\theta = \theta_{hkl}$) corresponds to planes with a normal direction, $\widehat{\mathbf{n}}_{hkl}$, given by

$$\widehat{\mathbf{n}}_{hkl} = \frac{\lambda}{4\pi \sin \theta_{hkl}} \left(h\mathbf{A} + k\mathbf{B} + l\mathbf{C} \right), \qquad (7.16)$$

where the scalar prefactor ensures unit length. This representation of the orientation of the planes, which is expressed in terms of the reciprocal vectors, can be transformed into a lattice-based picture by substituting it into eqn (7.13), the equation of the planes, with the general point r written as

$$\mathbf{r} = u\mathbf{a} + v\mathbf{b} + w\mathbf{c}. \qquad (7.17)$$

In conjunction with the orthogonality relations in eqn (7.8), it leads to the following condition on u, v and w for $\mathbf{r}$ to be on a plane perpendicular to $\widehat{\mathbf{n}}_{hkl}$:

$$\frac{\lambda}{2\sin\theta_{hkl}}\left(hu+kv+lw\right) = Nd_{hkl} + \Delta.$$

Thus the orientation of the atomic planes is such that they cut the a-axis (where $v=w=0$), the b-axis (with $u=w=0$) and the c-axis (when $u=v=0$) at

$$u = \frac{\eta}{h}, \quad v = \frac{\eta}{k} \quad \text{and} \quad w = \frac{\eta}{l}, \tag{7.18}$$

respectively, where η is a constant.

7.1.2 Simple consequences and applications

Before moving on, it is worth noting some elementary consequences and applications of Bragg's law. For example, the physical bound of $180°$ on scattering angle 2θ can be combined with eqn (7.15) to yield

$$\sin\theta = \frac{N\lambda}{2d} \leqslant 1.$$

With the integer N denoting the order of the reflections with respect to the straight-through beam (which is coincident with $N=0$), this shows that no Bragg peaks will be observed if $\lambda > 2d$. Too short a wavelength, when $\lambda \ll d$, might also be a problem as the large number of allowed reflections become difficult to distinguish as separate entities.

Although the determination of enough interplanar d-spacings can pin down the atomic positions in very simple materials, most samples require a much more detailed analysis of the diffraction pattern before their structure can be inferred. There is an engineering application, however, that relies primarily on the use of Bragg's law: the measurement of *residual stress*. The processes involved in the manufacture of mechanical components, such as railway tracks and turbine blades, often lock-in latent stresses that can contribute to their premature failure. These can be probed by examining the resulting distortions incurred by the crystal lattice. If a tensile stress σ causes the interplanar spacing to increase from a value of d_o in an annealed sample to d in the worked material, for example, then its magnitude can be estimated from the resultant *strain*, $(d-d_o)/d_o$, through a multiplication with the corresponding Young's modulus of elasticity, E:

$$\sigma = \left(\frac{d-d_o}{d_o}\right)E.$$

The most common 'direct' use of Bragg's law is not for d-spacing determination but wavelength selection. When a polychromatic beam of X-rays or neutrons impinges on a crystal of known structure and orientation, only those with specifically chosen wavelengths emerge at a given scattering angle of 2θ:

$$\lambda = \frac{2d\sin\theta}{N}\,.$$

While an ideal *monochromator* would select a single λ, the integer N in Bragg's law means that particles with a half, a third, a quarter, and so on, of the principal wavelength could also pass through. Most of these are unlikely to be present in the polychromatic source on energetic grounds but, if better monochromation is required, the emergent beam can be passed through a second crystal with a different d-spacing.

7.2 Patterns and symmetries

As mentioned earlier, the central feature of crystalline materials is their inherently repetitive nature. While they all share the translational symmetry of eqn (7.1), which gives rise to the reciprocal lattice, many display patterns within the unit cell. These additional symmetries in the SLD, in turn, lead to patterns in the intensities of the Bragg peaks.

7.2.1 Reality and Friedel pairs

A general property of $\beta(\mathbf{r})$ that we have assumed throughout this book, for simplicity, is the reality of the SLD:

$$\beta(\mathbf{r}) = \beta(\mathbf{r})^*. \tag{7.19}$$

As noted in eqn (2.47), its Fourier transform,

$$F(\mathbf{Q}) = \iiint\limits_V \beta(\mathbf{r})\,e^{i\mathbf{Q}\cdot\mathbf{r}}\,d^3\mathbf{r}\,, \tag{7.20}$$

is conjugate symmetric:

$$F(-\mathbf{Q}) = F(\mathbf{Q})^*. \tag{7.21}$$

For crystalline samples, $F(\mathbf{Q})$ can have non-zero values only when eqn (7.6) is satisfied. As such, eqn (7.21) becomes

$$F_{\bar{h}\bar{k}\bar{l}} = F_{hkl}^* \tag{7.22}$$

where $\bar{h} = -h$, $\bar{k} = -k$ and $\bar{l} = -l$. Since the measured signal scales with $|F|^2 = F\,F^*$, the intensity of the Bragg peak indexed by $h\,k\,l$ will be the same (subject to experimental corrections) as that for the reflection denoted by $\bar{h}\,\bar{k}\,\bar{l}$; these are known as *Friedel pairs*. A violation of Friedel's law is indicative of a breakdown of eqn (7.19), and is called *anomalous scattering*.

7.2.2 Centrosymmetry and reality

With a suitably chosen origin, which is arbitrary in the absence of a Fourier phase, the SLD function may be symmetric with respect to an inversion through $\mathbf{r} = 0$:

$$\beta(-\mathbf{r}) = \beta(\mathbf{r}) . \qquad (7.23)$$

Structures having this property are said to be *centrosymmetric*, and their Fourier transforms satisfy

$$F(-\mathbf{Q}) = F(\mathbf{Q}) . \qquad (7.24)$$

Assuming that the wavelength of the X-rays or neutrons is such that the crystalline sample contains no sources of anomalous scattering, eqn (7.24) can be combined with eqn (7.21) to yield

$$F_{hkl} = F_{hkl}^* . \qquad (7.25)$$

The structure factors of centrosymmetric crystals can be represented by real numbers, therefore, as their phases can be restricted to take values of only 0 or π radians.

7.2.3 Space groups and systematic absences

Just as the translational symmetry of eqn (7.1) enables the SLD of the entire crystalline sample to be deduced from a knowledge of the contents of a single unit cell, so too internal symmetries enable the SLD of the complete unit cell to be deduced from knowledge of a certain fraction of it known as the *asymmetric unit*. For the simple case of centrosymmetry, only one half of the unit cell is required:

$$0 \leqslant u \leqslant \tfrac{1}{2}, \quad 0 \leqslant v \leqslant 1, \quad 0 \leqslant w \leqslant 1,$$

for example, where u, v and w are *fractional* unit cell coordinates, as in eqn (7.17), so that lattice translation reduces to

$$\beta(u,v,w) = \beta(u+n_1, v+n_2, w+n_3) , \qquad (7.26)$$

where n_1, n_2 and n_3 are integers. The SLD in the rest of the unit cell, with $1/2 \leqslant u \leqslant 1$, is given by

$$\beta(u,v,w) = \beta(1-u, 1-v, 1-w) ,$$

where the locations parameterized by (u,v,w) and $(1-u, 1-v, 1-w)$ are said to be *equivalent positions*.

The number and type of equivalent positions define the symmetry of the crystal within its unit cell. There are 230 alternatives listed in Volume A of the *International Tables for Crystallography*, called *space groups*, and each leads to a corresponding set of relationships for its structure factors. Before illustrating this with an example, let us rewrite the Fourier transform of eqn (7.20) in a way that is better suited to crystalline samples; that is, in terms of *Miller* indices, hkl, and fractional unit cell coordinates. With $\mathbf{Q}$ and $\mathbf{r}$ given by eqns (7.6) and (7.17), and the orthogonality relations of eqn (7.8),

$$\mathbf{Q} \boldsymbol{\cdot} \mathbf{r} = 2\pi \left(hu + kv + lw \right) \quad \text{and} \quad \mathrm{d}^3\mathbf{r} = V_{\mathrm{cell}} \, \mathrm{d}u \, \mathrm{d}v \, \mathrm{d}w \, ,$$

so that eqn (7.20) becomes

$$\boxed{\;\mathrm{F}_{hkl} = V \int_0^1 \int_0^1 \int_0^1 \beta(u,v,w) \, \mathrm{e}^{\mathrm{i}2\pi(hu+kv+lw)} \, \mathrm{d}u \, \mathrm{d}v \, \mathrm{d}w\;} \, , \qquad (7.27)$$

following eqns (7.3) and (7.9).

Let us consider one of the most common space groups, $P\,2_1/c$, or number 14 in the International Tables, which has four equivalent positions:

$$
\begin{aligned}
(u,v,w) \, , \qquad &\left(1-u, \tfrac{1}{2}+v, \tfrac{1}{2}-w\right) , \\
(1-u, 1-v, 1-w) \, , \qquad &\left(u, \tfrac{1}{2}-v, \tfrac{1}{2}+w\right) ,
\end{aligned}
\qquad (7.28)
$$

with an asymmetric domain of

$$0 \leqslant u \leqslant 1, \quad 0 \leqslant v \leqslant \tfrac{1}{4}, \quad 0 \leqslant w \leqslant 1 \, .$$

According to the fundamental translational symmetry of eqn (7.26), any fractional coordinate that goes outside the range of zero to one on applying the equivalences of eqn (7.28) can be reassigned a location within the $[0,1]$ unit cell through the addition (or subtraction) of an appropriate integer. The Fourier transform of eqn (7.27) can then be evaluated through an integral over the asymmetric unit

$$\mathrm{F}_{hkl} = V \int_0^1 \int_{v=0}^{1/4} \int_0^1 \beta(u,v,w) \, \xi(h,k,l,u,v,w) \, \mathrm{d}u \, \mathrm{d}v \, \mathrm{d}w \, , \qquad (7.29)$$

where the function ξ is a sum of four complex exponentials, $\mathrm{e}^{\mathrm{i}\mathbf{Q}\boldsymbol{\cdot}\mathbf{r}}$, over the equivalent positions in eqn (7.28). After some trigonometric algebra, it is found that

$$\xi = 4 \cos\left[2\pi k v + \frac{\pi}{2}(k+l)\right] \cos\left[2\pi(hu+lw) - \frac{\pi}{2}(k+l)\right]. \qquad (7.30)$$

The fact that ξ is real, rather than complex, is a reflection of the centrosymmetry of the space group: the two equivalent positions in the second line of eqn (7.28) are related to those of the first by an inversion through the origin (and lattice translation). The other thing to note is that ξ, and hence F_{hkl}, will be equal to zero if either of the cosines in eqn (7.30) is zero. As this will occur when $k=0$ and l is odd, or when $h=l=0$ and k is odd, there will be a systematic absence

$$\mathrm{F}_{h0l} = 0 \ \text{ for } l \neq 2n$$
$$\mathrm{F}_{0k0} = 0 \ \text{ for } k \neq 2n$$

of certain Bragg peaks in the diffraction pattern. These are part of the fingerprint, in reciprocal space, of the symmetry within the unit cell. A more careful examination of the properties of eqn (7.30) leads to the conclusion that the structure factors come in related groups of four:

$$\mathrm{F}_{hkl} \;=\; \mathrm{F}_{\bar{h}k\bar{l}} \;=\; (-1)^{l+k}\,\mathrm{F}_{h\bar{k}l} \;=\; (-1)^{l+k}\,\mathrm{F}_{\bar{h}k\bar{l}}\,. \qquad (7.31)$$

The systematic absences, or *extinctions*, then follow as special cases. When $k=0$, for example,

$$\mathrm{F}_{h0l} \;=\; (-1)^{l}\,\mathrm{F}_{h0l} \;\Longrightarrow\; \mathrm{F}_{h0l} = 0 \ \text{ for } l = 2n+1\,,$$

where n is an integer. The theoretical equivalence of the intensity of groups of reflections, such as the quadruples in eqn (7.31), known as the *Laue symmetry*, is used to calibrate experimental corrections such as absorption.

7.2.4 Geometry and space groups

Although there is no mathematical constraint on the allowed SLD pattern within a unit cell, physical considerations, such as the quantum mechanical rules for atomic and molecular bonding, mean that only certain types of symmetry can occur in practice; this limits the number of potential space groups to 230. Indeed, a knowledge of the chemical content of a unit cell can restrict the possible space groups further. If the sample is *chiral*, for example, as might be expected for a biological compound, and enantiomerically pure, then the internal symmetry of the unit cell can only be consistent with the 65 space groups that preserve handedness.

As listed in the International Tables, each of the space groups is assigned a *crystal system*: triclinic, monoclinic, orthorhombic, tetragonal, trigonal, hexagonal and cubic. While these relate to the seven

Table 7.1 The seven possible geometrical shapes of a unit cell, as defined by their lattice parameters, and the space groups that share their characteristic symmetries (numbered in accordance with the *International Tables of Crystallography*).

Crystal system	Unit cell geometry	Space group numbers
triclinic	$a \neq b \neq c;\ \alpha \neq \beta \neq \gamma$	$1-2$
monoclinic	$a \neq b \neq c;\ \alpha = \gamma = 90° \neq \beta$	$3-15$
orthorhombic	$a \neq b \neq c;\ \alpha = \beta = \gamma = 90°$	$16-74$
tetragonal	$a = b \neq c;\ \alpha = \beta = \gamma = 90°$	$75-142$
trigonal	$a = b = c;\ \alpha = \beta = \gamma \neq 90°$	$143-167$
hexagonal	$a = b \neq c;\ \alpha = \beta = 90°, \gamma = 120°$	$168-194$
cubic	$a = b = c;\ \alpha = \beta = \gamma = 90°$	$195-230$

alternative shapes that a unit cell can have, depending on the relationships between the lengths of its sides and the angles between them, the association is one of shared symmetries rather than actual geometry. There is no formal requirement for a link between the internal SLD symmetry of a unit cell and its external morphology, since space groups can be specified in terms of equivalent positions that make no reference to lattice geometry, but they are found to be correlated strongly in practice. As such, the shape of the unit cell inferred from the reciprocal lattice provides the first indications about the probable space group of the crystal.

7.2.5 Symmetry and statistics

For reasons of atomic coordination, efficient molecular packing and energetic stability, crystallization does not occur with the same frequency in each of the 230 space groups. About a third of all small organic molecules conform to $P2_1/c$, for example, whereas proteins are found most often in the $P2_12_12_1$ and $P2_1$ configurations. Such statistical information can be helpful in the assessment of the space group, when the incomplete and noisy nature of experimental data leaves room for doubt on the basis of the Laue symmetry and systematic absences of the diffraction pattern.

An analysis of the distribution of Bragg peak intensities can also give an indication as to whether a crystal has adopted a centrosymmetric space group or a noncentrosymmetric one: the former tends to result in a higher proportion of reflections with very low intensity, I, relative to the average, μ, as compared to the latter. This was first noted by Wilson (1949), and the theoretical curves bearing his name that capture this tendency are plotted in Fig. 7.5. An alternative to the conventional explanation, which postulates randomly located atoms of comparable scattering length, is given in the appendix of Sivia and David (1994). There the two probability distributions are seen to represent our state of knowledge about the intensity of a reflection when given only the value of the average and the possibility of inversion symmetry.

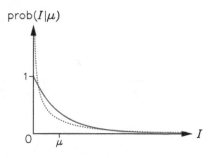

Fig. 7.5 The centric (dotted) and acentric (solid) Wilson distributions; $\langle I \rangle = \mu$.

7.3 Circumventing the phase problem

The biggest theoretical obstacle to ascertaining the structure of the crystal, $\beta_{\text{cell}}(\mathbf{r})$, from the diffraction data, $\{I_{hkl}\}$, is the lack of the Fourier phase:

$$I_{hkl} \propto |F_{hkl}|^2, \tag{7.32}$$

where the SLD in the unit cell is related linearly to F_{hkl} through eqn (7.27). The difficulties associated with the phase problem were highlighted in Section 2.6. Circumventing them hinges, ultimately, on our ability to supplement the missing information with suitable prior knowledge. This entails the use of chemical and physical knowledge about the sample to varying degrees but, in some circumstances, might also involve the collection of several sets of carefully chosen diffraction measurements.

7.3.1 Patterson maps

The easiest way of getting around the phase problem is to avoid it! Rather than trying to work with the SLD function, just consider its auto-correlation:

$$I_{hkl} \propto \int_0^1 \int_0^1 \int_0^1 \mathrm{ACF}\big[\beta(u,v,w)\big]\, e^{\mathrm{i}2\pi(hu+kv+lw)}\, \mathrm{d}u\, \mathrm{d}v\, \mathrm{d}w\,, \qquad (7.33)$$

where

$$\mathrm{ACF}\big[\beta(u,v,w)\big] = \int_0^1 \int_0^1 \int_0^1 \beta(x,y,z)^*\, \beta(x+u,y+v,z+w)\, \mathrm{d}x\, \mathrm{d}y\, \mathrm{d}z\,.$$

We discussed the ACF in a general one-dimensional context in Section 2.4.2, and in relation to the phase problem in reflectivity data in Section 4.1.4; we also met it in the guise of pair-correlation and pair-distribution functions in Sections 5.4 and 6.2. The main difference in the crystallographic case, where it is called a *Patterson map*, is that it's genuinely three-dimensional and often contains a high degree of symmetry. While the details of the latter depend on the space group of $\beta(\mathbf{r})$, it always has the translational symmetry of the crystal lattice:

$$\mathrm{ACF}(u,v,w) = \mathrm{ACF}(u+n_1, v+n_2, w+n_3)\,,$$

where n_1, n_2 and n_3 are integers. As two (point-like) atoms located at $\mathbf{r}_1$ and $\mathbf{r}_2$, and of scattering lengths b_1 and b_2 (assumed real), contribute to the ACF an equal signal, $b_1 b_2$, at $\mathbf{r}_1 - \mathbf{r}_2$ and $\mathbf{r}_2 - \mathbf{r}_1$,

$$\mathrm{ACF}\big[b_1\delta(\mathbf{r}-\mathbf{r}_1) + b_2\delta(\mathbf{r}-\mathbf{r}_2)\big] = \big(b_1^2 + b_2^2\big)\delta(\mathbf{r}) + b_1 b_2\, \delta\big(\mathbf{r}\pm(\mathbf{r}_1-\mathbf{r}_2)\big)\,,$$

Patterson maps are centrosymmetric.

The linearity of eqn (7.33) means that there is a one-to-one mapping between the diffraction pattern and the ACF of $\beta(\mathbf{r})$. As such, the loss of phase of F_{hkl} will be reflected in the Patterson map through its ambiguities with regard to the crystal structure. This is apparent from a simple consideration of their allowed symmetries: Patterson maps can adopt only 24 of the 230 space groups available to the SLD, implying that the relationship between them is not one-to-one. Despite this shortcoming, the real-space representation of the diffraction data provided by the ACF can be helpful in unravelling the crystal structure.

7.3.2 **Heavy atoms and partial structure**

Although it's difficult to ascertain $\beta(\mathbf{r})$ from $\mathrm{ACF}\big[\beta(\mathbf{r})\big]$ in general, the task of inferring the location and orientation of a certain fragment of the structure tends to be more accessible. If the sample is known to contain a heavy atom, for example, then the most intense peaks in the Patterson map from X-ray data are likely to be related to it; their analysis will often indicate its position. In fact, stoichiometry and space group symmetry alone can be enough to determine the location of some atoms. While the heavy atom scenario is rare in neutron work, as the strength of the nuclear interaction varies more uniformly across the periodic table, the presence of a negative scattering length among otherwise positive ones yields a Patterson map that is highly informative: the resultant negative peaks in the ACF will all be due to correlations between this reference atom and the rest of the structure. For macromolecules, such as proteins, the building blocks (amino acids) are usually well known. Their orientation may then be explored by looking for corresponding patterns of peaks in the Patterson map.

Suppose now that part of the crystal structure is 'known'. Denoting this generically as the heavy-atom fragment, $\beta_{\mathrm{H}}(\mathbf{r})$, and letting the unsolved portion be $\Delta\beta(\mathbf{r})$,

$$\beta(\mathbf{r}) = \beta_{\mathrm{H}}(\mathbf{r}) + \Delta\beta(\mathbf{r}) . \tag{7.34}$$

If F, $\mathrm{F_H}$ and $\Delta\mathrm{F}$ are the Fourier transforms of β, β_{H} and $\Delta\beta$, so that

$$\mathrm{F_H}(h,k,l) = V \int_0^1 \int_0^1 \int_0^1 \beta_{\mathrm{H}}(u,v,w)\, \mathrm{e}^{\mathrm{i}2\pi(hu+kv+lw)} \, \mathrm{d}u \,\mathrm{d}v \,\mathrm{d}w$$

and so on, then

$$\mathrm{F} = \mathrm{F_H} + \Delta\mathrm{F} \tag{7.35}$$

for any set of Miller indices. Of these three complex quantities, $\mathrm{F_H}$ can be calculated from $\beta_{\mathrm{H}}(\mathbf{r})$, the modulus of F can be estimated from the diffraction data and the expected value of $|\Delta\mathrm{F}|^2$ can be ascertained from the nature of the unlocated atoms (Wilson 1942):

$$|\mathrm{F}| = \mathrm{F_o} \pm \sigma \quad \text{and} \quad \big\langle |\Delta\mathrm{F}|^2 \big\rangle = \xi^2 .$$

If the fraction of scattering material in $\beta_{\mathrm{H}}(\mathbf{r})$ is sufficiently large, so that $|\mathrm{F_H}| > \xi$ (say), then the argument of $\mathrm{F_H}$, ϕ_{H}, will provide a good approximation to the phase of F:

$$\mathrm{F} \approx \mathrm{F_o} \exp(\mathrm{i}\phi_{\mathrm{H}}) . \tag{7.36}$$

The origin and reliability of this assignment can be understood from Fig. 7.6, which shows the constraint imposed on the value of F by a noisy measurement of its modulus-squared and information about

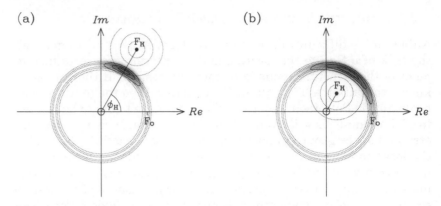

Fig. 7.6 The likelihood constraint on the value of a structure factor given a noisy measurement of its modulus-squared and a 'known' heavy-atom fragment.

the heavy-atom fragment. As discussed in Section 2.6.3, and illustrated in Fig. 2.23, the former restricts F to lie within a ring in the Argand plane. A knowledge of $\beta_H(\mathbf{r})$ and the average characteristics of $\Delta\beta(\mathbf{r})$, on the other hand, suggests that the most likely value of F is F_H, but that it could deviate from this isotropically, with decreasing probability, over a length scale of ξ. As shown in Fig. 7.6, the combination of these two constraints gives an indication of the likely phase of the structure factor. For reflections with well-determined intensities ($\sigma \ll \xi$), the reliability of ϕ_H as the phase of F is found to vary monotonically with

$$\text{FOM} = \frac{F_o\,|F_H|}{\xi^2}\,.$$

This 'figure-of-merit' was first noted by Woolfson (1956) and Sim (1960), and a modern example of its use can be found in Sivia and David (2001).

7.3.3 Isomorphous replacement

Through the selective addition or replacement of certain atoms in a crystal, it may be possible to produce two or more samples that differ from each other in a well-defined and structurally known manner. Denoting this calibrated change by $\Delta\beta(\mathbf{r})$, the original (or native) SLD, $\beta(\mathbf{r})$, and its *isomorphous* variant, $\beta_I(\mathbf{r})$, are related by

$$\beta(\mathbf{r}) = \beta_I(\mathbf{r}) + \Delta\beta(\mathbf{r})\,. \tag{7.37}$$

This is analogous to eqn (7.34), and its Fourier transform is similar to eqn (7.35):

$$F = F_I + \Delta F\,, \tag{7.38}$$

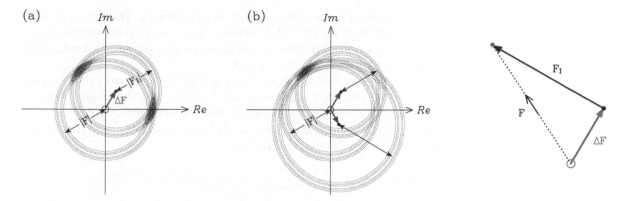

Fig. 7.7 (a) The likelihood constraint on the value of a structure factor, F, given additional diffraction data from (a) an isomorphous sample and (b) two such variants.

for any set of Miller indices. Unlike the heavy-atom case, however, ΔF can be calculated,

$$\Delta F(h,k,l) = V \int_0^1 \int_0^1 \int_0^1 \Delta\beta(u,v,w)\, e^{i2\pi(hu+kv+lw)}\, du\, dv\, dw,$$

and both $|F|$ and $|F_I|$ estimated from diffraction measurements on the native and isomorphous samples. The corresponding constraint imposed on the phase of F is illustrated in Fig. 7.7(a). Although the angular uncertainty is a fraction of the original 2π radians, there are two distinct possibilities. This ambiguity can be broken through the acquisition of scattering data from a further isomorphous variant, at least in principle, as shown in Fig. 7.7(b). The practical problem is that the derived samples are not usually as isomorphic as hoped: the addition of a heavy atom, for example, tends to distort the binding site in the native structure.

7.3.4 Direct methods and prior knowledge

Without the benefit of a heavy atom, or several sets of related data, the successful solution to the phase problem depends on the use of suitable prior knowledge. While many different $\beta(\mathbf{r})$ can give agreement with the diffraction intensities, $\{I_{hkl}\}$, relatively few (ideally just one) will be consistent with the chemical content of the sample and have the expected elemental bond lengths and angles. This information, and its implicit restriction on the structure factor phases, can be encoded through an appropriate analytical model for the SLD. In eqn (6.1), for example, $\beta(\mathbf{r})$ is described as the sum of N atoms of known type, and the structure is correspondingly defined by their locations. In the case of biological macromolecules, the connectivity of rigid molecular building blocks can often be exploited to express

$\beta(\mathbf{r})$ in terms of a modest number of parameters. Despite such efficiency, finding a set of model variables that yield agreement with the scattering measurements, in the absence of a good initial estimate, is a computationally difficult task.

Until the advent of silicon chips, a highly nonlinear optimization of the type involved above was not feasible. All that could reasonably be calculated was a direct Fourier inverse of the data, but this required the explicit assignment of phases to the experimentally derived structure factor amplitudes:

$$\beta(u, v, w) \approx \frac{1}{V} \sum_h \sum_k \sum_l |\mathrm{F}_{hkl}| \exp(\mathrm{i}\,\widehat{\phi}_{hkl})\, \mathrm{e}^{-\mathrm{i}2\pi(hu+kv+lw)}\,.$$

The crux of the problem, of course, is choosing the $\widehat{\phi}_{hkl}$! Theoretical progress on this issue was made by Karle and Hauptman (1950), for which they received the Nobel Prize in 1985. Starting with arbitrary values for two or three strong reflections, which fixes the origin of the unit cell, they were able to make predictions about other phases based on relationships that the structure factors had to obey since the electron density, $\beta(\mathbf{r})$ for X-rays, could not be negative. The analysis is complicated because the simple constraint of positivity in real space translates into a non-obvious set of inequalities in reciprocal space. The phase extension procedure is not definitive, however, and so a number of alternative assignments are usually explored. The ultimate test on the utility of this approach rests with the interpretation of the resulting SLD in terms of an atomic structure that makes chemical and physical sense. If it does, then the derived parameters can serve as a good point of entry for a model-based inference. Even when the $\beta(\mathbf{r})$ obtained in this manner is far from perfect, there may be enough clues in the noisy reconstruction to enable part of the structure to be solved; this information can then be used in the heavy-atom sense.

7.4 Powdered samples

The primary hurdle in crystallography is the actual growing of the sample. Producing a *single crystal* of sufficiently large size to yield good diffraction data is often time-consuming and a challenge, especially for neutron work. For this reason, scattering experiments are also conducted on *powdered* samples consisting of a very large ensemble of small randomly oriented crystallites. The lack of alignment means that, as in Fig. 5.7, the resulting differential cross-section is spherically symmetric:

$$\left(\frac{\mathrm{d}\sigma}{\mathrm{d}\Omega}\right)_{\mathrm{el}} \propto \mathrm{S}_{\mathrm{el}}(\mathbf{Q}) = \mathrm{S}_{\mathrm{el}}(Q)\,,$$

where $Q = |\mathbf{Q}|$. In terms of the geometry of Fig. 3.3, therefore, the diffraction pattern is invariant with respect to the azimuthal angle

(a)
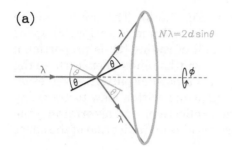
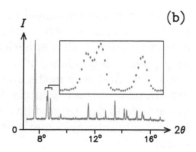
(b)

Fig. 7.8 (a) A Debye–Scherrer cone for a powdered sample. (b) Part of the diffraction pattern from a zeolite for monochromatic X-rays.

ϕ because the modulus of $\mathbf{Q}$ depends only on θ and the wavelength, λ. This leads to conical trajectories of equal scattering, as illustrated in Fig. 7.8(a), or circular rings of uniform intensity in a plane perpendicular to the incident beam.

An example of a powder pattern is shown in Fig. 7.8(b): it is a plot of the scattered intensity, I, as a function of 2θ for X-rays of a fixed wavelength. The peaks in the signal occur when the spacing between the atomic planes in the sample, d, and the angle θ satisfy Bragg's law of eqn (7.15). In terms of the reciprocal lattice formulation of eqn (7.6), this happens when

$$\frac{4\pi \sin\theta}{\lambda} = \left| h\mathbf{A} + k\mathbf{B} + l\mathbf{C} \right|. \tag{7.39}$$

Since this condition may be satisfied by more than one set of Miller indices, it leads to the problem that the area under a Bragg peak, $I(\sin\theta/\lambda)$, can correspond to a linear mixture of the structure factor intensities of several reflections:

$$I(\sin\theta/\lambda) \propto \sum_{(7.39)} \left| F_{hkl} \right|^2,$$

where the summation is over the hkl that obey eqn (7.39). The potential for such overlaps is easiest to see for a *cubic* unit cell, of side a, when eqn (7.39) reduces to the Pythagorean requirement

$$h^2 + k^2 + l^2 = \left(\frac{2a \sin\theta}{\lambda} \right)^2.$$

Simple coincidences, such as those due to a permutation of h, k and l above, pose no difficulty if the space group symmetry ensures the equality of the related $|F_{hkl}|$; the number of the latter is called the *multiplicity* of the reflection. The entanglement of the (005) set with (034) is a more troublesome cubic case, and representative of the diffraction information lost in having a powdered sample rather than a single crystal.

Peak overlap also occurs for reflections whose Qs are sufficiently close, though not identical, due to their non-zero widths; this can

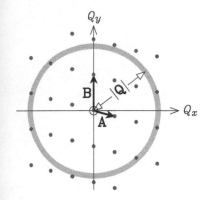

be seen from the expanded inset in Fig. 7.8(b). The severity of the problem increases with Q because the volume of reciprocal space enclosed within a thin spherical shell of radius $|\mathbf{Q}|$ is proportional to Q^2; the Q-dependence of the peak widths tends to exacerbate the effect. The first step in the analysis of a powder pattern involves an examination of the low-angle Bragg peaks, with a view to assigning hkl-indices to the isolated low-order reflections and ascertaining the lattice constants; systematic absences give an indication of the space group.

The spread and shape of the Bragg peaks results from a combination of instrumental factors, such as the source–sample–detector distances and the collimation angles, and the intrinsic properties of the sample. The size of the crystallites makes a contribution, for example, as the situation is akin to diffraction from a finite three-dimensional grating; the one-dimensional analogue is illustrated in Fig. 2.17. Residual stresses in the sample also smear out the peaks as the X-rays, or neutrons, see a distribution of d-spacing relative to the unstrained material.

Before moving on, we should note two other general features that can be seen in Fig. 7.8(b). Firstly, there is some scattered intensity between the Bragg peaks. This slowly varying signal can arise from several factors such as multiple scattering, poor shielding, incoherent (neutron) scattering and various types of aperiodic imperfections in the sample known as *disorder*. This background intensity must be subtracted, or accounted for, in order to carry out the crystalline analysis. Secondly, the Bragg peaks seem to diminish in magnitude with increasing θ. This is to be expected given the Q-dependence of X-ray form factors and the temperature-related Debye–Waller term, discussed in Sections 3.2.2 and 3.3.2.

7.4.1 Texture

The central assumption in powder diffraction is that of a very large number of randomly oriented crystallites. This starts to break down if there are too few of them, or if they are aligned to some degree. The latter is called *texture* or preferred orientation. A single crystal can be regarded as one extreme and an ideal powder as the other. Texture analysis is of interest in the intermediate regime found in worked materials and geological samples.

Texture can be quantified through an *orientational distribution function*, or ODF, which is usually specified in terms of *Euler angles*:

$$\mathrm{ODF} = \mathrm{ODF}(\Phi, \Theta, \Psi),$$

where $0 \leqslant \Phi \leqslant 360°$, $0 \leqslant \Theta \leqslant 180°$ and $0 \leqslant \Psi \leqslant 360°$ define the orientation of one set of Cartesian axes, $x\,y\,z$, taken to be aligned with a crystallite, relative to another, XYZ, assumed to be fixed within the sample as a whole. In essence, Φ and Θ correspond to the longitude and colatitude of the z-axis with respect to XYZ, with Z as north and X the prime meridian, and Ψ allows for a rotation of the $x–y$ plane about z. The fraction of crystallites with their Euler angles lying between Φ and $\Phi+\mathrm{d}\Phi$, Θ and $\Theta+\mathrm{d}\Theta$, Ψ and $\Psi+\mathrm{d}\Psi$ is given by the value of the ODF at (Φ,Θ,Ψ) times $\sin\Theta\,\mathrm{d}\Phi\,\mathrm{d}\Theta\,\mathrm{d}\Psi$, where $\sin\Theta$ is the relevant *Jacobian*. If $\mathrm{S}_{\mathrm{el}}(\mathbf{Q}|\Phi,\Theta,\Psi)$ is the scattering function of the crystal structure that has been rotated through Φ, Θ and Ψ, then the differential cross-section is its average over the ODF:

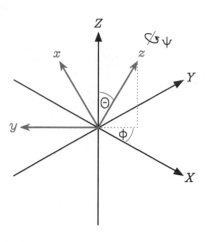

$$\left(\frac{\mathrm{d}\sigma}{\mathrm{d}\Omega}\right)_{\mathrm{el}} \propto \iiint \mathrm{S}_{\mathrm{el}}(\mathbf{Q}|\Phi,\Theta,\Psi)\ \mathrm{ODF}(\Phi,\Theta,\Psi)\ \sin\Theta\,\mathrm{d}\Phi\,\mathrm{d}\Theta\,\mathrm{d}\Psi.$$

In texture work, $\beta_{\mathrm{cell}}(\mathbf{r})$ is typically the known structure of a simple metal or mineral. This means that $\mathrm{S}_{\mathrm{el}}(\mathbf{Q}|\Phi,\Theta,\Psi)$ can be calculated, and the analysis task is one of trying to infer the ODF from the diffraction measurements.

7.4.2 Twinning

In single crystal work, a systematic departure from a δ-function ODF sometimes occurs when the internal symmetry of the unit cell is lower than that of its external geometry. The situation is illustrated schematically in Fig. 7.9, and is known as *twinning*. The differential cross-section in this case is a discrete sum of the two, or more, single crystal components comprising the sample:

$$\left(\frac{\mathrm{d}\sigma}{\mathrm{d}\Omega}\right)_{\mathrm{el}} = \sum_{j=1}^{few} \eta_j\ \mathrm{S}_{\mathrm{el}}(\mathbf{Q}|\Phi_j,\Theta_j,\Psi_j)\,,$$

where the domain coefficients $\{\eta_j\}$, and the crystal structure $\beta_{\mathrm{cell}}(\mathbf{r})$, are estimated from the diffraction data while the Euler angles are calculated for an assumed twinning symmetry.

A related problem involves the analysis of chiral structures when the sample contains enantiomerically pure regions of both left- and right-handedness. The coefficients η_{left} and η_{right} must then be combined with $\mathrm{S}_{\mathrm{el}}(\mathbf{Q}|\mathrm{left})$ and $\mathrm{S}_{\mathrm{el}}(\mathbf{Q}|\mathrm{right})$ to model the differential cross-section for a given $\beta_{\mathrm{cell}}(\mathbf{r})$.

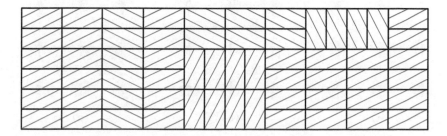

Fig. 7.9 A schematic illustration of (4-fold) twinning, whereby a sample consists of an uncorrelated mix of several types of symmetry-related single crystal domains.

7.4.3   Fibre diffraction

Some of the earliest molecular compounds to have their structures
solved were biological fibres; of these, the DNA work of Watson and
Crick (1953) is the most famous. Although the details depend on the
situation, diffraction from a fibrous sample is akin to the highly tex-
tured powder case where the long axis of all the crystallites is lined
up in a given direction. Rotational averaging about this unique axis,
z say, means that the differential cross-section is two-dimensional:

$$\left(\frac{d\sigma}{d\Omega}\right)_{el} \propto S_{el}(Q_r, Q_z),$$

where $Q_r^2 = Q_x^2 + Q_y^2$. While this is more informative than a powder
pattern, $S_{el}(Q)$, it is not as unambiguous as the three-dimensional
data from a single crystal. As such, fibre diffraction has largely become a niche technique.

7.5   Magnetic structures

Apart from a brief mention in Sections 1.4.3 and 3.3.4, and a length-
ier discussion in Section 3.2.3, we have implicitly taken the samples
to be non-magnetic. This is inconsequential for all but a few cutting-
edge X-ray experiments, but is a significant omission for neutrons.
Since a full theoretical treatment of magnetic scattering rapidly becomes very technical, we will restrict ourselves to a consideration of
some basic issues.

One of the reasons for the complexity of this topic is the vectorial nature of magnetic moments. Unlike nuclear scattering lengths,
they have both a magnitude and a direction. The analysis simplifies
considerably when the moments can only be in one of two directions,
'up' or 'down', for they can then be treated like scalars: they're just
positive or negative. Three crystalline structures of this type, corresponding to a binary compound, are illustrated in Fig. 7.10. With all

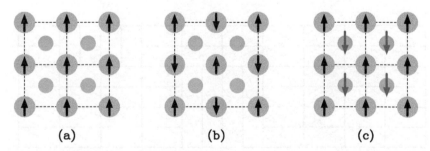

Fig. 7.10 Two-dimensional examples of (a) a ferromagnet, (b) an antiferromagnet
and (c) a ferrimagnet. The coloured discs represent the nuclear cross-sections and
the arrows indicate the magnetic moments.

its magnetic moments aligned, the first is a *ferromagnet*; the second, with alternating directions and no net moment, is an example of an *antiferromagnet*; the third, where the up and down contributions do not cancel completely, is a *ferrimagnet*.

Scattering from a repetitive pattern of moments leads to Bragg peaks in the differential cross-section, just as in the non-magnetic case. The nuclear and magnetic signals will overlap for (a) and (c) in Fig. 7.10, due to their shared unit cell, but there will be additional magnetic intensity at intervening points in (b) because the period of the moments is twice as large as that of the nuclear structure. When the temperature of the sample is raised sufficiently to destroy the ordering of the moments, thereby turning the material into a *paramagnet*, the magnetic portion of the scattering transitions into a slowly varying incoherent background. At very low temperatures, when the moments are ordered and the Debye–Waller term can be neglected, the magnetic peaks diminish with Q whereas the nuclear contribution continues out to high values.

Neutron diffraction data from a powdered inorganic compound are shown in Fig. 7.11 for two different temperatures: 5 K in blue and 295 K in grey. The peaks in the difference between them, which are shaded in light blue in the expanded inset, represent the Bragg signal from the magnetic structure. One of them coincides with a peak at room temperature, which is entirely from the nuclear structure, but the other does not; the 'new' intensity at $2\theta \approx 26°$ indicates that the sample is antiferromagnetic. The magnetic signal decays rapidly to zero at higher angles due to the associated form factor. The two sets of data are virtually identical beyond $2\theta \sim 30°$, therefore, apart from a slight shift in the Bragg positions due to the thermal expansion of the sample.

Magnetic moments can form repetitive patterns even when they are not restricted to lie in only one of two directions. If their period is a rational fraction of the size of the nuclear unit cell, such as the 2:1 in Fig. 7.10(b), then the two crystalline structures are said to be

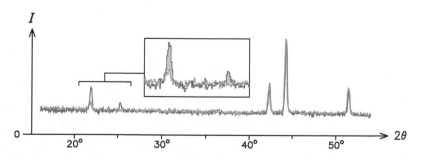

Fig. 7.11 Part of the low-angle neutron data from a powdered sample of the antiferromagnet $Co_{1.2}Fe_{0.8}Ge_{0.5}Ga_{0.5}Mo_3N$, taken on the D2b diffractometer at the ILL with $\lambda = 2.4$ Å, at a temperature of 5 K (blue) and 295 K (grey). (Sviridov *et. al.*, 2010)

commensurate; otherwise, they are incommensurate. The greater the complexity of the situation, the more the scattering experiments benefit from the use of polarized neutron beams and a single crystal sample. As the associated generalizations and analyses quickly lead into territory beyond the scope of this elementary text, the interested reader is referred to Bacon (1955), Squires (1978), Lovesey (1986), Lovesey and Collins (1996), and so on, for a more advanced and comprehensive treatment of the subject.

Part III

Inelastic scattering

Energy exchange and dynamical information

8

Having gained experience with elastic scattering, we now move on to consider the general case where there is an exchange of energy as well as momentum.

8.1 Experimental considerations

The first complication to arise when going from elastic to inelastic scattering is the need for two wavelengths, λ_i and λ_f for the incident and outgoing beams respectively, because

$$|\mathbf{k}_i| = \frac{2\pi}{\lambda_i} \neq \frac{2\pi}{\lambda_f} = |\mathbf{k}_f| \,. \tag{8.1}$$

The X-ray, or neutron, loses energy if $\lambda_i < \lambda_f$ and gains it if $\lambda_i > \lambda_f$. The wavevector diagram is shown in Fig. 8.1. Unlike in Fig. 3.2, the

$$\mathbf{Q} = \mathbf{k}_i - \mathbf{k}_f \tag{8.2}$$

triangle is not isosceles. Consequently, the moduli of these vectors and the scattering angle have to be related through the more general formula:

$$\boxed{Q^2 = k_i^2 + k_f^2 - 2\,k_i\,k_f\cos 2\theta} \;, \tag{8.3}$$

$k_i = |\mathbf{k}_i|$

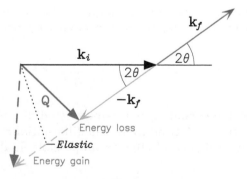

Fig. 8.1 The wavevector diagram for inelastic scattering through an angle of 2θ.

which reduces to eqn (3.7) in the elastic limit of eqn (3.6). Adopting the geometrical setup in Fig. 3.3, the vectorial content eqn (8.2) is encapsulated by

$$\mathbf{Q} = \frac{2\pi}{\lambda_f}\left(-\sin 2\theta \cos\phi, \ -\sin 2\theta \sin\phi, \ \frac{\lambda_f}{\lambda_i} - \cos 2\theta\right), \qquad (8.4)$$

which can be shown to satisfy eqn (8.3), as required, and to simplify to eqn (3.8) when $\lambda_i = \lambda_f$.

For a photon, the transition from λ_i to λ_f is associated with a frequency change of

$$\omega = 2\pi c\left(\frac{1}{\lambda_i} - \frac{1}{\lambda_f}\right) = \omega_i - \omega_f, \qquad (8.5)$$

where c is the speed of light, and a related transfer of energy

$$E = \hbar\omega = E_i - E_f \qquad (8.6)$$

to the sample. The corresponding expression for a neutron is derived from a consideration of its kinetic energy, $(\hbar k)^2/2m_\mathrm{n}$. In terms of the ω in eqn (8.6), it takes the form

$$\omega = \frac{\pi h}{m_\mathrm{n}}\left(\frac{1}{\lambda_i^2} - \frac{1}{\lambda_f^2}\right). \qquad (8.7)$$

The scattered particle loses energy if $E > 0$ and gains it if $E < 0$; the converse applies to the sample.

8.1.1   The partial differential cross-section

In Section 3.1, the differential cross-section, $\mathrm{d}\sigma/\mathrm{d}\Omega$, was introduced as the key object that summarized a scattering experiment. It was also noted that these $(2\theta, \phi)$-measurements could be translated into the more theoretical notion of a scattering law, $\mathrm{S}(\mathbf{Q})$, most easily if all the interactions were taken to be elastic and the incident beam was monochromatic. Assuming a unique λ_i, for simplicity, the idea can be extended to the inelastic case through the *partial differential cross-section*:

$$\frac{\mathrm{d}^2\sigma}{\mathrm{d}\Omega\,\mathrm{d}E'} = \lim_{\Delta E' \to 0}\left[\frac{\mathrm{d}\sigma/\mathrm{d}\Omega \text{ for } E_f \text{ between } E' \text{ and } E' + \Delta E'}{\Delta E'}\right] \qquad (8.8)$$

where, as indicated earlier, λ_f is related to E_f differently for X-rays and neutrons. In this formulation, the differential cross-section is given by the integral of eqn (8.8) over all possible final energies,

$$\frac{\mathrm{d}\sigma}{\mathrm{d}\Omega} = \int\limits_0^\infty \frac{\mathrm{d}^2\sigma}{\mathrm{d}\Omega\,\mathrm{d}E'}\,\mathrm{d}E', \qquad (8.9)$$

whereas the elastic component, which dominates the scattering, is just the value of the partial differential cross-section when there is no exchange of energy,

$$\left(\frac{d\sigma}{d\Omega}\right)_{el} = \left.\frac{d^2\sigma}{d\Omega\,dE'}\right|_{E'=E_i}. \tag{8.10}$$

Likewise, the cross-section, of Section 3.2.4, is equal to the integral of $d\sigma/d\Omega$ over all directions:

$$\sigma\left(\lambda_i\right) = \int_{2\theta=0}^{\pi}\int_{\phi=0}^{2\pi}\frac{d\sigma}{d\Omega}\,d\Omega\,, \tag{8.11}$$

where the element of solid angle $d\Omega = \sin 2\theta\,d2\theta\,d\phi$.

While the potential dependence of σ on λ_i has been made explicit, it is implicit in eqns (8.8) and (8.9). Indeed, the partial differential cross-section is additionally a function of 2θ, ϕ and λ_f, which is why it can be associated with $S(\mathbf{Q},\omega)$ through eqns (8.4)–(8.7). In terms of this scattering law,

$$S(\mathbf{Q}) = \int_{-\infty}^{\omega_i} S(\mathbf{Q},\omega)\,d\omega \quad \text{and} \quad S_{el}(\mathbf{Q}) = S(\mathbf{Q},\omega{=}0)\,, \tag{8.12}$$

where the incident particle can transfer no more than its initial energy, $\hbar\omega_i$, to the sample.

8.1.2 Triple-axis spectrometer

Following the work of Brockhouse (1955), inelastic experiments are traditionally carried out with a *triple-axis spectrometer*. Its main components, excluding the collimation hardware, are illustrated in Fig. 8.2. It differs from an elastic, or *double-axis*, setup through the addition of an analyser stage, since the final and incident energies are no longer assumed to be the same. The analyser is simply a monochromator that intercepts the scattered beam. Both selectively

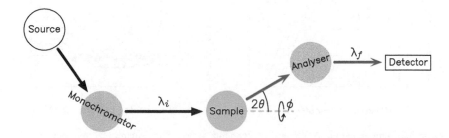

Fig. 8.2 A schematic illustration of a triple-axis spectrometer.

reflect particles of a given wavelength based on the Bragg condition satisfied by suitably oriented crystals, as discussed in Section 7.1.2. An appropriate choice of λ_i, λ_f, 2θ and ϕ allows the desired value of $\mathbf{Q}$ and ω to be accessed.

8.1.3 Time-of-flight instruments

Although a triple-axis spectrometer is very versatile, each monochromation stage takes a heavy toll on the particle flux. One of the two can be avoided naturally at a spallation neutron source by using the time-of-flight technique. The method was outlined for elastic scattering in Section 3.5.1, but the principle is the same in the inelastic case. If L_1 is the length of the flight path from the moderator to the sample, and L_2 is the distance from the sample to the detector, then the neutron will take a time

$$t = \frac{m_\mathrm{n}}{h}\left(L_1\lambda_i + L_2\lambda_f\right) \tag{8.13}$$

to travel from the source to the detector. This reduces to eqn (3.52) when $\lambda_i = \lambda_f$ and $L = L_1 + L_2$. With $t = 0$ set by the proton pulse of the accelerator, eqn (8.13) enables λ_f to be calculated from a measurement of the neutron's arrival time if λ_i is known; similarly, $\lambda_i = \lambda_i(t, \lambda_f)$. There is no need to have both a monochromator and an analyser, therefore, as either will suffice.

Time-of-flight setups where the incident wavelength is fixed are known as *direct geometry* instruments; they often employ mechanical *choppers* to select neutrons moving with a particular velocity. Those with a polychromatic incident beam and energy analysers for the scattered neutrons are said to have an *indirect geometry*. As a function of time, t, at a given 2θ and ϕ, such instruments explore $(\mathbf{Q},\omega)$-space along one-dimensional trajectories; this is illustrated in Fig. 8.3 for $|\mathbf{Q}|$ and E.

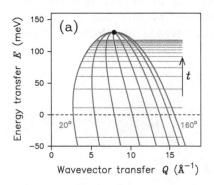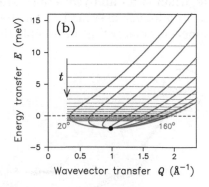

Fig. 8.3 The energy and momentum transfer accessed by various 2θ detectors in a time-of-flight instrument with (a) direct geometry ($E_i = 130\,\mathrm{meV}$, $L_1 = 12\,\mathrm{m}$, $L_2 = 4\,\mathrm{m}$) and (b) indirect geometry ($E_f = 2\,\mathrm{meV}$, $L_1 = 36\,\mathrm{m}$, $L_2 = 1.5\,\mathrm{m}$).

8.2 Scattering from time-varying structures

In Chapter 3, we first considered scattering from a single fixed atom and then moved on to an assembly of such entities. It was asserted that the time-invariant nature of the associated SLD, $\beta(\mathbf{r})$, guaranteed that the scattering interactions were elastic. A deeper insight into this claim emerges when we remove the temporal restriction, so that $\beta = \beta(\mathbf{r}, t)$, and generalize the analysis of Section 3.3 to the inelastic case.

With $\beta = \beta(\mathbf{r}, t)$, the time dependence of the incident and scattered waves must be made explicit. As such, eqn (3.12) becomes

$$\psi_i = \psi_\mathrm{o}\, e^{i(k_i z - \omega_i t)},$$

where ψ_o is a complex constant related to the incoming flux, Φ. Similarly, eqn (3.29) takes the form

$$\Phi \propto |\psi_\mathrm{o}|^2$$

$$\mathrm{d}\psi_f = \psi_\mathrm{o}\, e^{i(\mathbf{k}_i \bullet \mathbf{r} - \omega_i t)} \left[\beta(\mathbf{r}, t)\, \mathrm{d}^3\mathbf{r}\, \mathrm{d}t\right] \frac{e^{i[\mathbf{k}_f \bullet (\mathbf{r}' - \mathbf{r}) - \omega_f t]}}{|\mathbf{r}' - \mathbf{r}|} \qquad (8.14)$$

for the isotropic contribution to the outgoing wave at $\mathbf{r}'$ made by the SLD within an infinitesimally small volume centred on $\mathbf{r}$ between t and $t + \mathrm{d}t$. In the far-field limit, when $|\mathbf{r}' - \mathbf{r}| \to |\mathbf{r}'| = r'$, the integral of eqn (8.14) over the sample and time gives

$$\psi_f = \frac{\psi_\mathrm{o}\, e^{i\mathbf{k}_f \bullet \mathbf{r}'}}{r'} \iiint\limits_{V} \int\limits_{\tau} \beta(\mathbf{r}, t)\, e^{i(\mathbf{Q} \bullet \mathbf{r} - \omega t)}\, \mathrm{d}^3\mathbf{r}\, \mathrm{d}t\,,$$

where $\hbar\omega$ and $\hbar\mathbf{Q}$ are the energy and momentum transfers in the process. Since the probability that an incident particle will be scattered through 2θ and ϕ with a wavelength of λ_f is proportional to $|\psi_f|^2$, the partial differential cross-section is given by

$$\delta\Omega = \frac{\delta A}{r'^2}$$

$$\boxed{\frac{\mathrm{d}^2\sigma}{\mathrm{d}\Omega\, \mathrm{d}E'} \propto \frac{c_f}{c_i} \left| \iiint\limits_{V} \int\limits_{\tau} \beta(\mathbf{r}, t)\, e^{i(\mathbf{Q} \bullet \mathbf{r} - \omega t)}\, \mathrm{d}^3\mathbf{r}\, \mathrm{d}t \right|^2}\,, \qquad (8.15)$$

where $E' = \hbar\omega_f$ and

$$\frac{c_f}{c_i} = \begin{cases} \lambda_i/\lambda_f & \text{for neutrons,} \\ 1 & \text{for photons.} \end{cases} \qquad (8.16)$$

The ratio of final-to-initial speeds, c_f/c_i, enters because the incident and scattered flux depend in part on how fast the particles are moving. Being unity for both neutrons and X-rays in the elastic case, this term was omitted earlier for simplicity. In essence, the partial differential cross-section is the modulus-squared of the space–time Fourier transform of $\beta(\mathbf{r}, t)$; the latter is formally $S(\mathbf{Q}, \omega)$.

$$\frac{\mathrm{d}^2\sigma}{\mathrm{d}\Omega\, \mathrm{d}E'} \propto \frac{c_f}{c_i}\, S(\mathbf{Q}, \omega)$$

8.2.1   Elastic scattering

Having derived eqn (8.15), we can confirm that a static SLD does indeed give rise to purely elastic scattering. If $\beta(\mathbf{r}, t) = \beta(\mathbf{r})$, then the four-dimensional Fourier transform separates into a product of spatial and temporal terms,

$$\iiint_V \beta(\mathbf{r})\, \mathrm{e}^{\mathrm{i}\mathbf{Q}\cdot\mathbf{r}}\, \mathrm{d}^3\mathbf{r} \int_\tau \mathrm{e}^{-\mathrm{i}\omega t}\, \mathrm{d}t \, .$$

When integrated over a sufficiently long time, the latter is approximated well by a δ-function:

$$\int_{-\infty}^{\infty} \mathrm{e}^{-\mathrm{i}\omega t}\, \mathrm{d}t \;=\; 2\pi\, \delta(\omega) \, .$$

As this is non-zero only for $\omega = 0$, or $E' = E_i$, the interactions must be entirely elastic.

Equation (8.15) also helps us to understand that the elastic component of the scattering tells us about the *time-averaged structure* of the sample. To see this, we just need to substitute $\omega = 0$ and carry out the t-integral first:

$$\left.\frac{\mathrm{d}^2\sigma}{\mathrm{d}\Omega\, \mathrm{d}E'}\right|_{E' = E_i} \propto \left| \iiint_V \langle \beta(\mathbf{r}, t) \rangle_\tau\, \mathrm{e}^{\mathrm{i}\mathbf{Q}\cdot\mathbf{r}}\, \mathrm{d}^3\mathbf{r} \right|^2 , \qquad (8.17)$$

where

$$\langle \beta(\mathbf{r}, t) \rangle_\tau \;=\; \frac{1}{\tau} \int_0^\tau \beta(\mathbf{r}, t)\, \mathrm{d}t \, . \qquad (8.18)$$

Information about the *dynamical* aspects of the SLD is contained in the inelastic part of the scattering function.

8.2.2   Space–time correlation function

The presence of the modulus-squared in eqn (8.15) makes it very difficult for $\beta(\mathbf{r}, t)$ to be inferred from the partial differential cross-section, in general. As noted on several occasions, however, the underlying phase problem is avoided if we work instead with the ACF,

$$G(\mathbf{r}, t) \;=\; \iiint_V \int_\tau \beta(\mathbf{R}, t')^*\, \beta(\mathbf{r}+\mathbf{R}, t+t')\, \mathrm{d}^3\mathbf{R}\, \mathrm{d}t', \qquad (8.19)$$

because this is related linearly to the scattering function:

$$\frac{\mathrm{d}^2\sigma}{\mathrm{d}\Omega\, \mathrm{d}E'} \propto \frac{c_f}{c_i} \underbrace{\iiiint_V \int_\tau G(\mathbf{r}, t)\, \mathrm{e}^{\mathrm{i}(\mathbf{Q}\cdot\mathbf{r} - \omega t)}\, \mathrm{d}^3\mathbf{r}\, \mathrm{d}t}_{S(\mathbf{Q}, \omega)} \, . \qquad (8.20)$$

Apart from being four-dimensional, $G(\mathbf{r}, t)$ is essentially no different from the ACFs met earlier: it gives us a measure of the correlation between pairs of points in the sample which are separated by a displacement r and a time t. Simultaneously occurring spatial associations are encoded in $G(\mathbf{r}, 0)$, therefore, and temporal linkages at the same location are captured by $G(0, t)$. If a structure is static, then $G(\mathbf{r}, t) = G(\mathbf{r})$.

Scattering theory was first developed from the perspective of the space–time correlation function by Van Hove (1954). In terms of this, the results of Section 8.2.1 are

$$\frac{d^2\sigma}{d\Omega \, dE'} \propto \delta(\omega) \iiint_V G(\mathbf{r}) \, e^{i\mathbf{Q}\bullet\mathbf{r}} \, d^3\mathbf{r} \qquad (8.21)$$

for a static structure and

$$\left(\frac{d\sigma}{d\Omega}\right)_{el} \propto \iiint_V \langle G(\mathbf{r}, t) \rangle_\tau \, e^{i\mathbf{Q}\bullet\mathbf{r}} \, d^3\mathbf{r} \,, \qquad (8.22)$$

with G replacing β in eqn (8.18).

8.2.3 Total scattering

As noted in Section 3.1.3, most diffraction experiments do not have any energy transfer discrimination and rely on the preponderance of elastic scattering:

$$\frac{d\sigma}{d\Omega} = \left(\frac{d\sigma}{d\Omega}\right)_{el} + \left(\frac{d\sigma}{d\Omega}\right)_{inel} \approx \left(\frac{d\sigma}{d\Omega}\right)_{el}.$$

Using the $G(\mathbf{r}, t)$ formulation, we are in a position to appreciate the structural significance of the quantity that is actually measured: the total differential cross-section. Substituting for $d^2\sigma/d\Omega \, dE'$ from eqn (8.20) in eqn (8.9),

$$\frac{d\sigma}{d\Omega} \propto \iiint_V \int_\tau G(\mathbf{r}, t) \, e^{i\mathbf{Q}\bullet\mathbf{r}} \, f(t) \, d^3\mathbf{r} \, dt \qquad (8.23)$$

where

$$f(t) = \int_0^\infty \frac{c_f}{c_i} \, e^{-i\omega t} \, dE' \qquad (8.24)$$

is the function resulting from the integral of everything that could depend on E'. With $c_f = c_i$ for photons, and $\hbar\omega = E_i - E'$,

$$f(t) = \hbar \int_{-\infty}^{\omega_i} e^{-i\omega t} \, d\omega \propto \delta(t) \,. \qquad \hbar \, d\omega = -dE'$$

Putting this δ-function into eqn (8.23), the t-integral becomes very easy and leads to

$$\frac{\mathrm{d}\sigma}{\mathrm{d}\Omega} \propto \iiint_{V} \mathrm{G}(\mathbf{r},0)\, \mathrm{e}^{\mathrm{i}\mathbf{Q}\bullet\mathbf{r}}\, \mathrm{d}^{3}\mathbf{r} \qquad (8.25)$$

for X-rays. The same expression holds for neutrons, albeit with a different proportionality constant, but showing it requires extra effort. It hinges on the fact that, to within a scaling factor,

$$\mathrm{f}(t) = \hbar \int_{-\infty}^{\omega_i} \sqrt{\frac{\omega_i - \omega}{\omega_i}}\, \mathrm{e}^{-\mathrm{i}\omega t}\, \mathrm{d}\omega$$

is also approximated well by a δ-function at $t=0$.

The distinction between the elastic and total differential cross-sections follows from eqns (8.22) and (8.25): the former is related to the time-averaged structure whereas the latter pertains to an instantaneous view, or snap-shot, of it. The two will be identical, of course, if $\mathrm{G}(\mathbf{r},t) = \mathrm{G}(\mathbf{r})$. While this might be a reasonable assumption for a solid, it's far from being applicable for a liquid sample. In fact, liquids should produce no elastic scattering, other than at $\mathbf{Q}=0$, because their time-averaged SLD is uniform. The analysis of such data depends crucially on the 'static approximation', whereby $|\omega| \ll \omega_i$, as the pseudo-elastic condition of $\lambda_i \approx \lambda_f$ is necessary, in the absence of $\mathrm{d}^2\sigma/\mathrm{d}\Omega\,\mathrm{d}E'$, for transforming the measured $\mathrm{d}\sigma/\mathrm{d}\Omega$ into the corresponding $\mathrm{S}(\mathbf{Q})$.

8.2.4 Coherent and incoherent scattering

In Section 3.3.3, we saw how the elastic differential cross-section for the nuclear scattering of neutrons could be expressed as the sum of a coherent and an incoherent contribution. A similar decomposition can be performed for inelastic scattering:

$$\frac{\mathrm{d}^2\sigma}{\mathrm{d}\Omega\,\mathrm{d}E'} = \left(\frac{\mathrm{d}^2\sigma}{\mathrm{d}\Omega\,\mathrm{d}E'}\right)_{\mathrm{coh}} + \left(\frac{\mathrm{d}^2\sigma}{\mathrm{d}\Omega\,\mathrm{d}E'}\right)_{\mathrm{incoh}}. \qquad (8.26)$$

In Chapter 3, the coherent component arose from the part of the SLD that showed a discernible spatial pattern and the incoherence resulted from the random nature of what was left over. With the extra time dimension associated with inelastic scattering, $(\mathrm{d}^2\sigma/\mathrm{d}\Omega\,\mathrm{d}E')_{\mathrm{coh}}$ is related to the correlated motion of the atoms in the sample while $(\mathrm{d}^2\sigma/\mathrm{d}\Omega\,\mathrm{d}E')_{\mathrm{incoh}}$ has to do with their independent behaviour. Thus coherent scattering tells us about collective excitation modes, such as *phonons* and *magnons*, whereas incoherent scattering contains information on local dynamics, such as *diffusion* of single atoms. We will see examples of both in Chapter 9.

8.3 A quantum transitions approach

Other than invoking wave–particle duality for the neutron and X-ray probes, through use of the de Broglie and Planck hypotheses, and appealing to some basic spin arithmetic when necessary, our analysis has been based entirely on classical physics. We have not taken into account the quantum properties of the sample itself in any meaningful way. The remedy for this shortcoming rests on a reformulation of the scattering problem in terms of transitions between discrete initial and final states of the particle–sample system as a whole. Since a proper treatment of the subject rapidly becomes quite daunting, even within the usual approximation of 'first-order time-dependent perturbation theory', we will merely highlight a few key points. Details can be found in more advanced texts on scattering theory, such as Lovesey (1986), and quantum mechanics books, such as Baym (1969).

As before, the incident and outgoing energy and momentum of the scattered particle is determined by the wavevectors $\mathbf{k}_i$ and $\mathbf{k}_f$. The difference now is that the corresponding initial and final state of the sample, indexed by η and η' respectively, must also be specified to define the scattering event completely. Using Dirac's compact notation, and ignoring issues of spin and polarization for simplicity, transitions from the configuration $|\mathbf{k}_i, \eta\rangle$ to $|\mathbf{k}_f, \eta'\rangle$ will occur at a rate of

$$\mathcal{R}_{\mathbf{k}_i, \eta \to \mathbf{k}_f, \eta'} = \frac{2\pi}{\hbar} \left| \langle \mathbf{k}_f, \eta' | \mathcal{V} | \mathbf{k}_i, \eta \rangle \right|^2 , \qquad (8.27)$$

subject to the conservation of energy constraint that

$$\left| \hbar\omega + E_\eta - E_{\eta'} \right| \lesssim \frac{h}{\Delta t} , \qquad (8.28)$$

$$\omega = \omega_i - \omega_f$$

where $\mathcal{V}(\mathbf{r})$ is the perturbative Fermi *pseudopotential*, which acts for a time Δt, representing the interaction between the sample and the X-ray or neutron. The number of transitions will also depend on the probability of finding the the sample in the initial state $|\eta\rangle$, P_η. Assuming a thermal equilibrium at temperature T, the latter will follow a Boltzmann distribution:

$$P_\eta = \text{prob}\Big(|\eta\rangle \Big| T\Big) \propto \exp\left(-\frac{E_\eta}{k_{\mathrm{B}}T}\right). \qquad (8.29)$$

As scattering measurements relate only to changes in the wavevector of the probe, the relevant transition rate is given by a weighted summation over all the possible sample states:

$$\mathcal{R}_{\mathbf{k}_i \to \mathbf{k}_f} = \sum_\eta P_\eta \sum_{\eta'} \mathcal{R}_{\mathbf{k}_i, \eta \to \mathbf{k}_f, \eta'} \; \delta\left(\hbar\omega + E_\eta - E_{\eta'}\right), \qquad (8.30)$$

where the constraint of eqn (8.28) has been replaced by a δ-function on the understanding that $h/\Delta t \approx 0$.

The partial differential cross-section requires an evaluation of the rate, $R(2\theta, \phi, E')$, at which particles are scattered into a tiny solid angle, $d\Omega$, at $(2\theta, \phi)$ with E_f between E' and $E' + dE'$. This entails a summation of eqn (8.30) over the small volume of *phase space*,

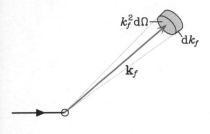

$$d^3\mathbf{k}_f = k_f^2 \, d\Omega \, dk_f = k_f^2 \, d\Omega \, \frac{dk_f}{dE_f} \, dE_f, \tag{8.31}$$

which is consistent with the given direction and final energy range. Assuming that $\mathcal{R}_{\mathbf{k}_i \to \mathbf{k}_f}$ does not change abruptly with $\mathbf{k}_f$, a uniform continuum of possibilities for the outgoing wavevector implies that

$$R(2\theta, \phi, E') \propto \mathcal{R}_{\mathbf{k}_i \to \mathbf{k}'} \, d^3\mathbf{k}'. \tag{8.32}$$

Substituting from eqns (8.31) and (8.32) into the basic definition

$$\frac{d^2\sigma}{d\Omega \, dE'} = \frac{R(2\theta, \phi, E')}{N \, \Phi \, d\Omega \, dE'},$$

where the incident flux $\Phi \propto c_i$ and the normalizing N is the number of scattering units in the sample, and using the fact that

$$\frac{dE_f}{dk_f} = \hbar \, c_f$$

for both X-rays and neutrons, we find that

$$\frac{d^2\sigma}{d\Omega \, dE'} \propto \frac{k'^2}{c_i \, c'} \mathcal{R}_{\mathbf{k}_i \to \mathbf{k}'} \tag{8.33}$$

where $\mathcal{R}_{\mathbf{k}_i \to \mathbf{k}'}$ follows from eqns (8.27) and (8.30). This is the usual starting point of more erudite discussions on scattering theory, but we will not pursue this approach further here.

Examples of inelastic scattering

In this final chapter, we consider several examples of inelastic scattering. Our aim here is not to be comprehensive, or too theoretical, but to gain an appreciation of the types of dynamical behaviour that can be probed with X-rays and neutrons.

9.1 Compton scattering

The earliest example of inelastic scattering at the atomic level dates back to 1922, with the work of Compton on the scattering of X-rays by a thin foil of graphite. He found that the signal at any given scattering angle, 2θ, contained two wavelengths: one was equal to that of the incident monochromatic beam, λ_i, as expected from classical radiation theory, but the other was longer. The wavelength shift of the latter, $\delta\lambda = \lambda_{2\theta} - \lambda_i$, depended on 2θ according to

$$\delta\lambda = \frac{h}{m_{\mathrm{e}}c}\left(1 - \cos 2\theta\right), \qquad (9.1)$$

and the process was associated with the emission of electrons. Compton explained this component of the scattering by considering it to be the result of a billiard ball type of collision between a photon and a stationary electron, as illustrated in Fig. 9.1.

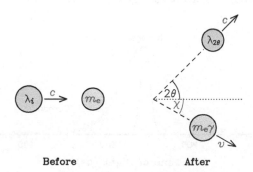

Before   **After**

$$\gamma = \left(1 - \frac{v^2}{c^2}\right)^{-1/2}$$

Fig. 9.1 The kinematical picture used by Compton to explain the unexpected change in the wavelength of some X-rays upon scattering from an electron. The relativistic correction is encoded through the introduction of the Lorentz factor, γ.

The historical importance of Compton's work was that it provided support for a particle picture of photons at a time when the notion was still controversial: they could be associated with a well-defined momentum, h/λ, and energy, hc/λ, and interact with an individual electron. While Compton's analysis was based on a collision with a stationary electron, a relaxation of this condition shows that information on the momenta of the electrons can be obtained from such experiments. Since the X-ray case is complicated by the need to use *relativistic mechanics*, as significant Compton scattering only occurs at incident energies comparable to $m_e c^2$, let us consider instead the collision between a neutron and a nucleus; the physical principles are the same.

If the momentum of a nucleus of mass M before the collision is $\mathbf{p}_i$, and that after the impact by the neutron is $\mathbf{p}_f$, then this two-body interaction must satisfy

$$\mathbf{p}_f - \mathbf{p}_i = \hbar\mathbf{k}_i - \hbar\mathbf{k}_f = \hbar\mathbf{Q}, \qquad (9.2)$$

from the *conservation of momentum*, where $\mathbf{k}_i$ and $\mathbf{k}_f$ are the usual incident and final wavevectors of the neutron, and

$$\frac{|\mathbf{p}_f|^2}{M} - \frac{|\mathbf{p}_i|^2}{M} = \frac{|\hbar\mathbf{k}_i|^2}{m_n} - \frac{|\hbar\mathbf{k}_f|^2}{m_n} = 2\hbar\omega, \qquad (9.3)$$

from the *conservation of energy* (in the classical limit). Substituting for $\mathbf{p}_f$ from eqn (9.2) into eqn (9.3),

$$|\mathbf{p}_f|^2 = \mathbf{p}_f \cdot \mathbf{p}_f = (\hbar\mathbf{Q} + \mathbf{p}_i) \cdot (\hbar\mathbf{Q} + \mathbf{p}_i),$$

we obtain

$$\mathbf{Q} \cdot \mathbf{p}_i = M\omega - \frac{\hbar|\mathbf{Q}|^2}{2}. \qquad (9.4)$$

Hence the strength of the scattered neutron signal, with an energy and momentum transfer of $\hbar\omega$ and $\hbar\mathbf{Q}$, will be related to the number of nuclei with a $\mathbf{p}_i$ that satisfy eqn (9.4). If all the nuclei were

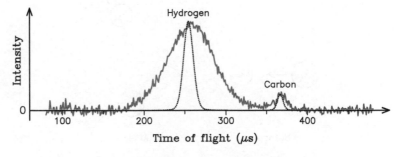

Fig. 9.2 Neutron data from polythene, at a scattering angle of $2\theta = 50°$, taken on the VESUVIO spectrometer at the ISIS facility. The expected signal from stationary nuclei of hydrogen and carbon is indicated with a dotted line. (Courtesy of Dr. J. Mayers.)

stationary, so that $\mathbf{p}_i = 0$, then there would be a sharp peak in the energy transfer spectrum whenever

$$E = \hbar\omega = \frac{\hbar^2 |\mathbf{Q}|^2}{2M} . \tag{9.5}$$

For a fixed $|\mathbf{Q}| = Q$, this means there will be as many peaks as nuclear species (masses) in the sample. They will be smeared out somewhat due to the instrumental resolution function, of course, but any additional broadening would be indicative of a spread in the initial momenta. An example of such data, from a sample containing only carbon and hydrogen, are shown in Fig. 9.2.

Focusing on the scattering from a single nuclear species, typically hydrogen, the preceding discussion can be summarized mathematically by

$$S_M(\mathbf{Q}, \omega) \propto \iiint n(\mathbf{p}) \, \delta\left(\omega - \frac{\hbar Q^2}{2M} - \frac{\mathbf{Q} \cdot \mathbf{p}}{M}\right) d^3\mathbf{p} , \tag{9.6}$$

where $n(\mathbf{p})$ is the *momentum distribution* of the atoms of mass M within the sample and the value of the δ-function is nought unless its argument is equal to zero. For the stationary case, $n(\mathbf{p}) = \delta(\mathbf{p})$. Given the central requirement of eqn (9.4) for observing a Compton signal, the related scattering function is often expressed in terms of the unit vector $\hat{\mathbf{Q}} = \mathbf{Q}/Q$ and the component of $\mathbf{p}$ in that direction,

$$y = \mathbf{p} \cdot \hat{\mathbf{Q}} = \frac{M}{Q}\left(\omega - \frac{\hbar Q^2}{2M}\right), \tag{9.7}$$

instead of $\mathbf{Q}$ and ω. Formally,

$$J_M(\hat{\mathbf{Q}}, y) = \frac{\hbar Q}{M} S_M(\mathbf{Q}, \omega) . \tag{9.8}$$

The advantage of this change of variables, or 'y-scaling', is that the *Compton profile* becomes one-dimensional when $n(\mathbf{p}) = n(|\mathbf{p}|)$:

$$J_M(\hat{\mathbf{Q}}, y) = J_M(y)$$

because the scattering cannot depend on the direction of $\mathbf{Q}$ when the momentum distribution is isotropic.

9.1.1 The impulse approximation

Perhaps the strangest thing about Compton scattering is not that the incident X-ray or neutron can be treated like a particle, but that the entity struck behaves as though it were an isolated object rather than an integral part of the structure within which it resides. That is the assumption, at least, under which the analysis is carried out. It is based on the *impulse approximation*, whereby the collisional interaction takes place so quickly that the neighbouring atoms, or electrons, don't have time to respond to the changing situation. Its validity depends on the incident energy and momentum transferred being sufficiently high that the local binding energy is negligible and the related interaction time scale is much shorter than that of any process in the sample.

The uncorrelated nature of the collision process above means that Compton scattering is incoherent. With reference to the neutron case and Section 3.2.4, the strength of the signal from a particular nuclear species then depends its average cross-section, $4\pi \langle b^2 \rangle$. In terms of the definitions in eqns (3.26) and (3.27), this is actually the sum of the coherent and incoherent cross-sections; as such, it is usually called σ_{tot}. The much larger size of the hydrogen peak in Fig. 9.2 compared to that of carbon, even though their number ratio in polythene ($-CH_2-$) is only two-to-one, is due to its enormous incoherent cross-section. For X-rays, the scattering length involved is the *Compton wavelength*, λ_C, rather then the classical (Thomson) radius of the electron, r_e, of eqn (3.19). Their ratio is

$$\lambda_C = \frac{h}{m_e c}$$

$$\frac{r_e}{\lambda_C} = \frac{e^2}{4\pi \epsilon_o hc} = \frac{\alpha}{2\pi} \approx \frac{1}{861} \, , \qquad (9.9)$$

where α is the *fine structure constant*.

9.1.2   Single particle wave function

The momentum distributions at the heart of Compton scattering are also linked to other quantities. The simplest connection is between an $n(\mathbf{p})$ and the average kinetic energy of the associated particle. For nuclei of mass M,

$$\langle \text{K.E.} \rangle = \frac{\langle |\mathbf{p}|^2 \rangle}{2M} = \frac{1}{2M} \iiint |\mathbf{p}|^2 \, n(\mathbf{p}) \, d^3\mathbf{p} \, , \qquad (9.10)$$

where we have assumed that $n(\mathbf{p})$ is correctly normalized so that its integral is equal to unity. For an isotropic distribution, this is related to the fourth moment of $n(p)$ and, somewhat less obviously, the second moment of $J(y)$.

At a more fundamental level, the *wave function*, ψ, describing the quantum mechanical behaviour of a particle can be expressed in terms of its position, $\mathbf{r}$, or its momentum $\mathbf{p}$. The two are connected through a Fourier transform:

$$\psi(\mathbf{p}) = h^{-3/2} \iiint \psi(\mathbf{r}) \, e^{i\mathbf{p} \cdot \mathbf{r}/\hbar} \, d^3\mathbf{r} \, , \qquad (9.11)$$

where $n(\mathbf{p}) = |\psi(\mathbf{p})|^2$ and the spatial probability density is given by $|\psi(\mathbf{r})|^2$. The relationship between $n(\mathbf{p})$ and $\psi(\mathbf{r})$, in turn, provides a link between Compton scattering and the potential, $\mathcal{V}(\mathbf{r})$, experienced by the associated particles, through *Schrödinger's equation*. An interesting example can be found in Reiter *et al.* (2002).

9.2 Lattice vibrations

In sharp contrast to the incoherent nature of Compton scattering, X-rays and neutrons can also interact with the correlated motions of the atoms within a sample. They can create and annihilate collective vibrational modes, such as sounds waves, called *phonons*. While this behaviour does occur in liquids and amorphous solids, it is seen most clearly in crystalline materials.

The simplest physical model that allows for an understanding of phonons consists of a long linear chain of atoms, all of mass m and separation d, connected by identical springs of stiffness α. This is illustrated in Fig. 9.3, with u_n denoting the displacement of the nth atom from its equilibrium position. Applying Newton's *second law of motion* ($F = ma$) to atom n,

$$\alpha\,(u_{n+1} - u_n) - \alpha\,(u_n - u_{n-1}) = m\,\frac{\mathrm{d}^2 u_n}{\mathrm{d}t^2} \qquad (9.12)$$

where we have assumed that the displacements are small enough that *Hooke's law* is obeyed and t is time. Following Sections 2.1 and 2.2, the pattern of us will constitute a travelling wave, of frequency ω and wavenumber k, if

$$u_j = A\,\mathrm{e}^{\mathrm{i}(k\,x_j - \omega t)}, \qquad (9.13)$$

for $j = 1, 2, 3, \ldots, \sim 10^{23}$, where $x_j = jd$ and $|A| \ll d$. Substituting this into eqn (9.12), with $j = n-1$, n and $n+1$, it can be verified as a valid solution provided that

$$\omega^2 = \frac{4\alpha}{m}\,\sin^2\!\left(\frac{kd}{2}\right). \qquad (9.14)$$

Taking the frequency to be positive, but allowing the wavenumber to be negative (for a 'backwards' travelling wave), the dispersion curve has the property

$$\omega(k) = \omega(-k) \qquad \text{and} \qquad \omega(k) = \omega\!\left(k + \frac{2\pi}{d}\right). \qquad (9.15)$$

This symmetry and periodicity means that $\omega(k)$ is defined uniquely for $0 \leqslant k \leqslant \pi/d$. The region $|k| \leqslant \pi/d$ is known as the first *Brillouin zone*.

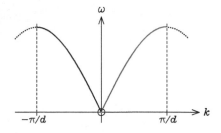

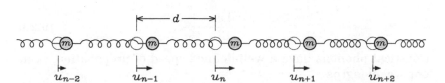

Fig. 9.3 A one-dimensional model that exhibits the most basic phonon behaviour.

If the one-dimensional ball-and-spring model is modified to have two different masses, m and M, alternating along the chain, then it leads to an interesting new behaviour also characteristic of phonons: there can be more than one type of lattice vibration, with its own frequency, for any given wavenumber. Inserting the atoms of mass M in between the existing ones in Fig. 9.3, to maintain the periodicity over d, and denoting their displacements from equilibrium by v_j, for $j = 1, 2, 3, \ldots, \sim 10^{23}$, the travelling wave

$$u_j = A\,\mathrm{e}^{\mathrm{i}(k\,x_j - \omega t)} \quad \text{and} \quad v_j = B\,\mathrm{e}^{\mathrm{i}(k\,x_j - \omega t)}, \tag{9.16}$$

where $x_j = j\,d$, can be shown to satisfy the *equations of motion* provided that

$$\omega^2 = \alpha\left(\tfrac{1}{m} + \tfrac{1}{M}\right) \pm \alpha\sqrt{\left(\tfrac{1}{m} + \tfrac{1}{M}\right)^2 - \tfrac{4}{mM}\sin^2\!\left(\tfrac{kd}{2}\right)}. \tag{9.17}$$

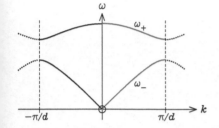

Although this expression is more complicated than eqn (9.14), the two solutions, $\omega_+(k)$ and $\omega_-(k)$, share the symmetries of eqn (9.15). The positive and negative branches of the dispersion curve are called the *optical* and *acoustical* modes, respectively. The origin of these terms is best understood by considering the nature of the associated vibrations in the limit of long wavelengths. For $|k| \ll \pi/d$, eqn (9.17) reduces to

$$\omega_+^2 \approx 2\alpha\left(\frac{1}{m} + \frac{1}{M}\right) \quad \text{and} \quad \omega_-^2 \approx \frac{\alpha\,k^2 d^2}{2\,(m+M)}. \tag{9.18}$$

The corresponding ratio of amplitudes in eqn (9.16) is found to be

$$\left[\frac{A}{B}\right]_+ \approx -\frac{M}{m} \quad \text{and} \quad \left[\frac{A}{B}\right]_- \approx 1, \tag{9.19}$$

from the equations of motion. Neighbouring atoms move in opposite directions for the positive solution, therefore, but in harmony for the negative one. If the two types of atoms carried equal and opposite charges, as for a binary ionic compound, the ω_+ mode would result in a fluctuating electric field; since this can couple to 'light', it is known as the optical branch. The acoustical mode is so called because the atomic motions are similar to those in a sound wave: they give rise to alternating regions of compression and rarefaction but not optical activity, as there is no net separation of ionic charge. Being non-dispersive for $\lambda \gg d$, when

$$\frac{\omega_-}{k} = \frac{\mathrm{d}\omega_-}{\mathrm{d}k} = c, \tag{9.20}$$

acoustical phonons have a well-defined speed of propagation, c, at long wavelengths.

For simplicity, the displacement of the atoms has so far been taken as collinear with the chain. As such, the discussion has been about *longitudinal* phonons. Vibrational modes in which the u_js and v_js

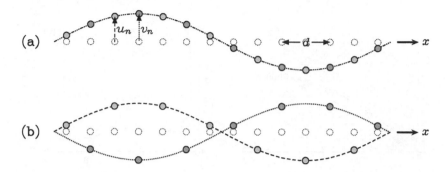

Fig. 9.4 Instantaneous atomic displacements, over a wavelength ($\lambda = 7d$), for transverse phonons along a linear binary chain: (a) is for an acoustical mode, and (b) is its optical counterpart.

are perpendicular to the direction of propagation, x, are said to be *transverse* phonons. They exhibit very similar behaviour in terms of their dispersion curves, including optical and acoustical characteristics; the latter are illustrated in Fig. 9.4. There are two transverse modes for every longitudinal excitation, since there are two *linearly independent* orientations that are perpendicular to the direction of propagation.

While the one-dimensional model above may be crude, it captures many of the salient features of phonons. The main difference from real samples, which are three-dimensional, is that the dispersion curves become a function of the wavevector k. For a crystalline material with lattice vectors a, b and c, the symmetries of eqn (9.15) generalize to

$$\omega(\mathbf{k}) = \omega(-\mathbf{k}) \quad \text{and} \quad \omega(\mathbf{k}) = \omega(\mathbf{k}+\mathbf{G}), \qquad (9.21)$$

where $\mathbf{G}$ is a linear integer combination of the reciprocal vectors $\mathbf{A}$, $\mathbf{B}$ and $\mathbf{C}$ defined in eqn (7.7).

$$\mathbf{G} = h\mathbf{A} + k\mathbf{B} + l\mathbf{C}$$

Figure 9.5 shows the neutron scattering function, $S(\mathbf{Q}, E)$, from a single crystal sample of copper selenide, for $\mathbf{Q}$ along the 110 direction. The magnitude of the wavevector transfer is given in *reciprocal lattice units*, whereby the boundary of the first Brillouin zone is at ± 0.5. The greatest intensity, by far, is in the neighbourhood of the Bragg peak at $(0,0)$, as indicated by the (logarithmic) blue shading. A weak plume-like signal, highlighted by the low-level contours, is also apparent and resembles the dispersion curve of an acoustical phonon. The latter arises because $S(\mathbf{Q}, E)$ records the creation and annihilation of phonons. That is to say, neutrons transfer an energy $\hbar\omega$ and momentum $\hbar \mathbf{k}$ to the sample when a phonon of frequency ω and wavevector k is generated; E and $\mathbf{Q}$ are negative to the same extent when it is destroyed. Taking into account the symmetry of eqn (9.21), the interaction is summarized by

$$E = \pm\hbar\omega \quad \text{and} \quad \mathbf{Q} = \mathbf{G} \pm \mathbf{k}, \qquad (9.22)$$

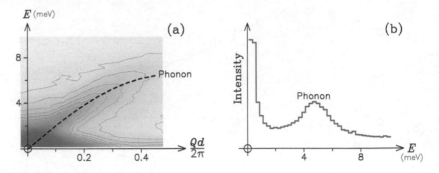

Fig. 9.5 (a) Neutron scattering data, $S(\mathbf{Q}, E)$ for momentum transfer along the 110 direction, from the superionic conductor $Cu_{1.8}Se$, taken on the TAIPAN spectrometer at ANSTO. (b) A plot of the scattered intensity, as a function of energy transfer, for $Q = 0.35$ reciprocal lattice units. (Courtesy of Drs. S. A. Danilkin and M. Yethiraj.)

where k is restricted to lie in the first Brillouin zone.

Phonons have traditionally been studied with neutron scattering, since their energy at atomic wavelengths is comparable to that of lattice vibrations. The technical requirements to attain similar measurements with X-rays are much more demanding, as there is only a tiny fractional change in the energy and momentum of the photon. The first example of a dispersion curve using X-rays was due to Dorner *et al.* (1987), some thirty years after the earliest work with neutrons. The sheer intensity of the latest generation of synchrotron sources, however, is starting to make X-rays a viable alternative for exploring lattice dynamics.

9.2.1   Heat capacities

The phonons studied microscopically with X-ray and neutron scattering also manifest themselves in certain macroscopic properties of the sample. We alluded to one of these in eqn (9.20), where the initial slope of the dispersion curve for an acoustical mode gives the 'speed of sound' in the material. Another link occurs in *heat capacities*. Following standard texts such as Blakemore (1985) and Kittel (1953), the phonon contribution to the internal energy, U, at a temperature T is given by

$$U = \int_0^\infty \frac{g(\omega)\,\hbar\omega}{\exp(\hbar\omega/k_B T) - 1}\,\mathrm{d}\omega\,, \tag{9.23}$$

where k_B is the Boltzmann constant and $g(\omega)\,\mathrm{d}\omega$ is the number of lattice vibrational modes with a frequency between ω and $\omega + \mathrm{d}\omega$. The *density of states*, $g(\omega)$, can either be inferred from heat capacity measurements, $\partial U/\partial T$, for a series of different temperatures, or from scattering data. As $T \to 0$, when the *Bose–Einstein* occupancy term and Debye–Waller factor can be ignored, $g(\omega)$ reduces to the in-

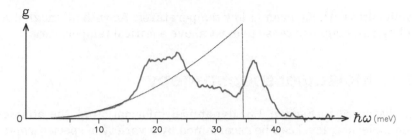

Fig. 9.6 The density of phonon states in aluminium, $g(\omega)$, estimated from neutron scattering measurements taken on the MARI spectrometer at the ISIS facility; the dotted line is the *Debye model*. (Courtesy of Prof. S. M. Bennington.)

tegral of $S(\mathbf{Q}, \omega)$ with respect to $\mathbf{Q}$ over the phonon peaks. A density of states function derived from neutron scattering measurements is shown in Fig. 9.6.

9.2.2 Spin waves

If a crystalline sample is magnetic, then the spins of the unpaired electrons can oscillate in a correlated fashion analogous to phonons; the excitations are called *spin waves* or *magnons*. The phenomenon may be visualized by considering a simple one-dimensional model: a set of bar magnets that are spaced uniformly and fixed along the x-axis, but whose north and south poles are free to rotate in the yz-plane. In the lowest energy state, neighbouring magnets will be aligned antiparallel to each other. If the 'left-most' bar of this chain is rotated about the x-axis, then the others will follow in an attempt to restore the equilibrium of the antiferromagnetic order. A plot of the y- or z-component of the magnetic moments will be a sinusoidal function of x and time, similar to Fig. 9.4(b), resulting from an angular lag, $\Delta\theta$, between successive bars. A dependence of $\Delta\theta$ on the frequency of rotation, ω, means that the x-distance over which the helical pattern of bar magnet orientations repeats itself, or the wavelength λ, will satisfy a characteristic dispersion relation: $\omega = \omega(k)$.

As the frequency of rotation in the mechanical model above is reduced, we would expect λ to become longer because neighbouring bar magnets will find it easier to keep up with each other. This intuition, that $k \to 0$ as $\omega \to 0$, is supported by a more formal analysis which predicts that

$$\omega \propto \begin{cases} k^2 & \text{for a ferromagnet,} \\ |k| & \text{for an antiferromagnet,} \end{cases} \tag{9.24}$$

for long wavelength magnons, and has been confirmed by neutron scattering experiments. Due to the magnetic form factor of Section 3.2.3, the intensity of $S(\mathbf{Q}, \omega)$ along a spin wave dispersion curve

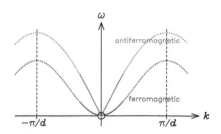

diminishes with $|\mathbf{Q}|$ even at low temperatures. As with all magnetic behaviour, magnons cease to exist above a critical temperature.

9.3 Molecular spectroscopy

As discussed in Section 1.3, dynamical information at the atomic and molecular level can be ascertained by a variety of spectroscopic techniques. X-rays extend the energy range that can be probed with optical and UV radiation, and enable the deep electronic structure of heavier elements to be studied. Thermal neutrons provide an alternative to IR, Raman and microwave methods for exploring molecular vibrations and rotations, as indicated by the similarity of the spectra in Fig. 9.7, but what benefit is there to using them? After all, access to neutron facilities is limited by comparison with IR and Raman instruments.

Neutrons possess qualities that offer some advantages over traditional electromagnetic techniques. One of these is the absence of *selection rules* for the observable transitions. As in the case of the optical mode in phonons, IR and Raman techniques are only sensitive to certain types of molecular motions; neutrons, on the other hand, can exchange energy and momentum with all kinds of excitations. This means that all modes are, in principle, observable by inelastic neutron scattering (INS) spectroscopy. In practice, the

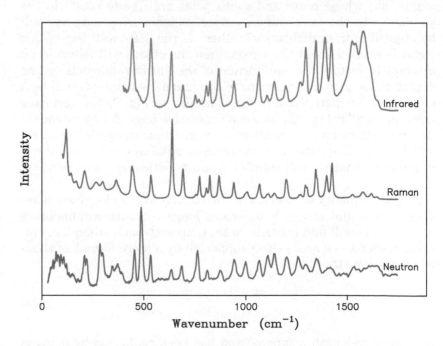

Fig. 9.7 Part of the infrared (IR), Raman and inelastic neutron scattering (INS) spectra of the α-amino acid L-cysteine. (Parker and Haris, 2008)

scattered intensity is dominated by modes that involve hydrogen displacement. The highly penetrating nature of neutrons is also an important property, especially when dealing with samples that are opaque to photons. It enables the study of adsorbed species on oxide and metal catalysts, for example, and hydrogen-in-metals. Another positive feature of neutrons is that the intensity of the scattered signal, $S(\mathbf{Q}, \omega)$, can be predicted fairly easily for a given molecular model. The corresponding calculation for IR and Raman work is much more difficult, and so agreement with the measurements is limited to a comparison of the *normal mode* frequencies. While the latter help to pin down the interatomic 'spring constants', an understanding of the intensity of the signal allows for an inference of the amplitudes of the motions.

Figure 9.8 shows a schematic plot of neutron scattering data from a molecular crystal, integrated over $\mathbf{Q}$, for positive energy transfers. The signal for $E < 0$ would be a mirror reflection with a diminishing intensity. This is due to the Boltzmann factor, whereby lower energy states in the sample are more likely to be occupied than higher ones. As a result, it is more probable that the neutron will lose energy in the encounter than to gain it; the ratio between the two possibilities is called *detailed balance*.

9.3.1 Quasi-elastic scattering

The most unusual feature in Fig. 9.8 is the broad line at $E = 0$. Being coincident with the elastic component, it is known as *quasi-elastic* scattering. An insight into its origin can be gained by recalling the Fourier relationship between the time-dependent structure, $\beta(\mathbf{r}, t)$, and the scattering function, $S(\mathbf{Q}, \omega)$, of eqn (8.15). Whereas a lack of temporal variation leads to a sharp signal at $\omega = 0$, and a sinusoidal oscillation at frequency ω_o yields peaks at $\pm \omega_o$, stochastic changes over a characteristic time-scale τ give rise to a line centred on $\omega = 0$ with a width dependent on $1/\tau$. The last assertion reduces to the elastic case in the limit of $\tau \to \infty$, as expected, but it may seem more

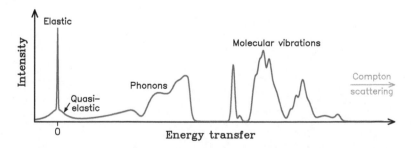

Fig. 9.8 A schematic plot of neutron scattering data from a hypothetical molecular crystal, $S(\mathbf{Q}, \omega)$ integrated over $\mathbf{Q}$, based on Eckert (1992).

plausible given a Fourier result of the type

$$\left| \int_0^\infty e^{-t/\tau}\, e^{i\omega t}\, \mathrm{d}t \right|^2 = \frac{1}{\omega^2 + (1/\tau)^2}\,, \tag{9.25}$$

where the function on the right is called a Lorentzian. Leaving the details to Bée (1988), an analysis of quasi-elastic scattering provides information on the diffusive translational and reorientational motion of atoms and molecules.

9.3.2 Energy resolution and time-scales

The central equations of Section 8.2, (8.15) and (8.20), indicate that there is a reciprocal Fourier relationship between ω and t just as there is between $\mathbf{Q}$ and r. The result given in eqn (9.25) is an explicit example. The resolution with which the energy transfer $E = \hbar\omega$ can be measured, ΔE, therefore has a bearing on the time-scale τ of the dynamics that can be explored:

$$\tau \sim \frac{1}{\Delta\omega} = \frac{\hbar}{\Delta E}\,. \tag{9.26}$$

While rapid motions can be inferred from low-resolution energy data, slower ones require a very small ΔE. In the quasi-elastic region, the highest energy resolution is achieved with the *spin-echo* technique (Mezei 1972) which makes use of the magnetic dipole of the neutron. With a ΔE as little as a few neV, time-scales of up to 100 ns become accessible.

Discrete Fourier transforms

A

The discrete nature of computer technology means that the continuum formulation of Fourier transforms in Section 2.4 is best replaced with a digitized version for numerical calculations. In the simplest case, this takes the form

$$F_k = \sum_{j=0}^{N-1} f_j \, e^{-i2\pi jk/N},$$ (A.1)

where the integers j and $k = 0, 1, 2, \ldots, N-1$ and the f_j represent a *sampling* of the function $f(x)$ on a grid of N points that are spaced evenly along the x-axis, and

$$f_j = \frac{1}{N} \sum_{k=0}^{N-1} F_k \, e^{i2\pi jk/N}.$$ (A.2)

With the complementary complex exponentials in the summations, these clearly have the essential ingredients to be a Fourier pair. To see how they relate to the discussion in Chapter 2, we need to examine them more carefully.

The first thing to note is that both f_j and F_k are implicitly periodic, with a repeat unit of N *pixels*:

$$f_j = f_{j+N}$$ (A.3)

from eqn (A.2), since

$$e^{i2\pi(j+N)k/N} = e^{i2\pi jk/N} e^{i2\pi k} = e^{i2\pi jk/N},$$

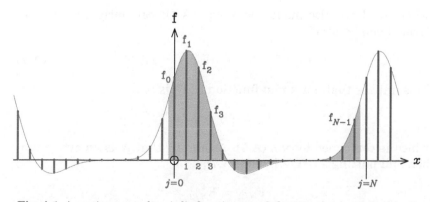

Fig. A.1 A continuous and periodic function sampled at N points per repeat unit.

as illustrated in Fig. A.1; likewise,

$$F_k = F_{k+N} \tag{A.4}$$

from eqn (A.1).

Although the periodicity of f_j means that an independent interval of N consecutive pixels can be chosen anywhere along the x-axis, the location assigned to the point $j=0$ is highly significant: as in the continuum case, it defines the origin. For example, the substitution of a δ-function at $j=0$,

$$f_j = \begin{cases} 1 & \text{if } j=0, \\ 0 & \text{otherwise}, \end{cases}$$

into eqn (A.1) yields $F_k = 1$ for all k, as would be expected from eqn (2.46) when $f(x) = \delta(x)$. By contrast, a displacement of the non-zero pixel to $j \neq 0$ leads to the F_k being complex: a uniform modulus, but an argument that increases linearly with k. The corresponding position of the origin in reciprocal space is easiest to verify by putting $k=0$ in eqn (A.1). It gives

$$F_0 = \sum_{j=0}^{N-1} f_j, \tag{A.5}$$

which is the discrete analogue of eqn (2.49).

With the periodicity of eqn (A.3), and the origin at $j=0$, an even function is one that satisfies

$$f_j = f_{-j} = f_{N-j}.$$

Hence, the symmetry relationships of eqn (2.48) become

$$f_j = \begin{cases} f_{N-j} \\ -f_{N-j} \end{cases} \quad \Longleftrightarrow \quad F_k = \begin{cases} F_{N-k}, \\ -F_{N-k}, \end{cases} \tag{A.6}$$

where we have also made use of eqn (A.4). Similarly, the discrete form of eqn (2.48) is

$$f_j = f_j^* \quad \Longleftrightarrow \quad F_{N-k} = F_k^*. \tag{A.7}$$

This implies that, for a real function f_j, F_0 is real:

$$F_0 = F_N = F_0^*,$$

which is confirmed by eqn (A.5). Assuming that N is an even number, and putting $k = N/2$ in eqn (A.7),

$$F_{N/2} = F_{N/2}^*.$$

Thus the *Nyquist* component, $F_{N/2}$, must also be real.

A consideration of Fig. A.1 shows that the location of pixel j corresponds to a value of x given by

$$j \equiv x_j = \frac{x_{max}}{N}\, j\,, \tag{A.8}$$

where x_{max} is the length of the repeat unit. While x_{max} is chosen at our discretion, it has to be bigger than the (finite) extent of the continuous function, $f(x)$, for the calculation to be meaningful. If a convolution is to be performed with $g(x)$, say, then x_{max} needs to be larger than the combined spread of $f(x) \otimes g(x)$ to ensure that *aliasing* artefacts, resulting from the inherent periodicity of the sum, are avoided. The equivalence between the integer k and the related continuous variable, q, can also be expressed in the manner of eqn (A.8):

$$k \equiv q_k = \frac{q_{max}}{N}\, k\,, \tag{A.9}$$

where the reciprocity of Fourier length scales dictates that

$$q_{max} \propto \frac{N}{x_{max}}\,. \tag{A.10}$$

The constant of proportionality is 2π if the continuum exponential is of the form $\exp(\pm iqx)$; it's unity for $\exp(\pm i2\pi qx)$. Equation (A.10) requires qualification, in that q_{max} must be at least twice the highest 'measured' frequency: due to eqn (A.4), the Fourier coefficients pertaining to

$$|k| \leqslant N/2\,,$$

or $|q| \leqslant q_{max}/2$, define the F_k spectrum completely. According to eqn (A.7), only the components with $0 \leqslant k \leqslant N/2$ need be specified if the function f_j is real.

The discrete Fourier transform (DFT) of eqns (A.1) and (A.2) can be evaluated very efficiently by using a fast Fourier transform (FFT) algorithm. Whereas the direct computation of a DFT entails $\mathcal{O}(N^2)$ operations, because each of the F_k involves the sum of N products, the same result can be obtained with just $\mathcal{O}(N\log_2 N)$ calculations by using the clever factorization of an FFT. The associated gain in computational speed, of $\mathcal{O}(N/\log_2 N)$, is enormous even for modest N. The only thing to bear in mind is that, for the most common FFT algorithm (Cooley and Tukey, 1965), N has to be a power-of-two:

$$N = 2^M,$$

where M is a positive integer. This requirement can be fulfilled by 'padding out' the f_j function, or the F_k spectrum, with zeros in an appropriate fashion. A suitable procedure for the example of Fig. A.1 is illustrated in Fig. A.2: with the increment in x, Δx, unchanged between the j-pixels, $q_{max} = 1/\Delta x$ remains the same; the larger N, however, means that the F_k sample the Fourier spectrum much more finely in q.

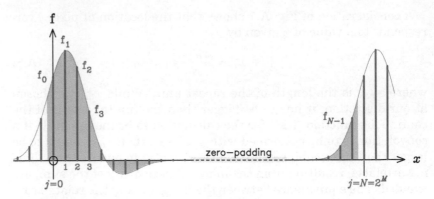

Fig. A.2 The function in Fig. A.1 zero-padded to make N a power-of-two.

The DFT of eqns (A.1) and (A.2), and its FFT implementation, can be generalized easily to many dimensions. Explicitly, the case of two takes the form

$$F_{kk'} = \sum_{j=0}^{N-1} \sum_{j'=0}^{N'-1} f_{jj'} \, e^{-i2\pi(jk/N + j'k'/N')} \tag{A.11}$$

and

$$f_{jj'} = \frac{1}{NN'} \sum_{k=0}^{N-1} \sum_{k'=0}^{N'-1} F_{kk'} \, e^{i2\pi(jk/N + j'k'/N')} \tag{A.12}$$

where the integers j and j' relate to x and y, k and k' correspond to q_x and q_y, and the digitization is on a grid of $N \times N'$.

Resonant scattering and absorption

For simplicity, the X-ray form factors, neutron scattering lengths and SLDs were taken to be real throughout the book. While this is a good approximation for thermal neutrons in most cases, and often appropriate for X-rays, the imaginary component cannot be ignored in the vicinity of certain wavelengths. These correspond to energies at which an electron from a core shell can be ejected by an X-ray photon, or a compound nucleus may be created with an additional neutron; as such, they are associated with a high rate of X-ray and neutron absorption.

The variation of $f(\lambda, \theta)$ of eqn (3.13) in the neighbourhood of one of these special energies, $\hbar\omega_o$, can be described by the addition of a complex term that mimics the behaviour of an isotropic *damped driven harmonic oscillator* close to resonance. That is to say,

$$f(\lambda, \theta) = f_o(\lambda, \theta) + f_R(\lambda) \qquad \text{(B.1)}$$

where $f_o(\lambda, \theta)$ has the characteristics discussed in Section 3.2 and $f_R(\lambda)$ is the resonant contribution. Pursuing the classical analogy,

$$\frac{d^2 x}{dt^2} + \eta\frac{dx}{dt} + \omega_o^2 x = e^{i\omega t} \qquad \text{(B.2)}$$

is the the one-dimensional equation of motion for a harmonic oscillator, of natural frequency ω_o, being driven by a sinusoidal force in a viscous medium ($\eta > 0$). The amplitude of the steady-state response, $x = A\,e^{i\omega t}$, is easily shown to be

$$A = \frac{1}{\omega_o^2 - \omega^2 + i\eta\omega}.$$

Identifying $A(\omega)$ with $f_R(\lambda)$, subject to a scaling constant and the relevant transformation between frequency and wavelength,

$$f_R(\lambda) \propto \frac{\omega_o^2 - \omega^2 - i\eta\omega}{\left(\omega_o^2 - \omega^2\right)^2 + \eta^2\omega^2}. \qquad \text{(B.3)}$$

This is plotted in Fig. B.1.

Although an $f(\lambda, \theta)$ with a non-zero imaginary component leads to a breakdown of Friedel's law of Section 7.2.1, the strong variation of a scattering length with λ close to a resonance can be exploited to

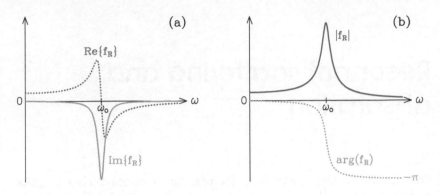

Fig. B.1 The resonant contribution of eqn (B.3), $f_R(\lambda)$, for $\eta = 0.1\omega_o$.

circumvent the phase problem. The method is similar to that of Section 7.3.3 in principle, and known as *multi-wavelength anomalous diffraction* (or MAD); details may be found in Hendrickson (1991) and Materlik *et al.* (1994).

The absorption cross-section of eqn (3.28), σ_{abs}, turns out to be proportional to $\mathcal{I}m\{f_R(\lambda)\}$. This is best understood by considering the undeflected passage of an X-ray or neutron ($\theta = 0$), of wavelength λ *in vacuo*, through a uniform medium of SLD

$$\beta = \frac{\rho N_A}{m}\left[f_o(\lambda, 0) + f_R(\lambda)\right],\qquad (B.4)$$

where $\rho N_A/m$ is simply the atomic number density. Using the refractive index relations of eqns (4.25) and (4.36), the wavenumber inside the material is given by

$$k = \frac{2\pi}{\lambda} - \beta\lambda.$$

The intensity of the wave, $\psi = \psi_o\,e^{ikx}$, will then diminish exponentially with distance l,

$$I = |\psi|^2 = I_o\,e^{2\mathcal{I}m\{\beta\}\lambda l},$$

if $\mathcal{I}m\{\beta\} < 0$. A comparison between this equation and (3.28), along with (B.4), yields

$$\sigma_{abs} = -2\lambda\,\mathcal{I}m\{f_R(\lambda)\},\qquad (B.5)$$

where we have implicitly assumed that $\mathcal{I}m\{f_o(\lambda, \theta)\} = 0$.

References

[1] *International Tables for Crystallography*. First online edition (2006) http://it.iucr.org/ (IUCr).

[2] Abelès, F. (1948). Sur la propagation des ondes electromagnetiques dans les milieux stratifies. *Ann. de Phys.*, **3**, 504–20.

[3] Als-Nielsen, J. and McMorrow, D. (2001). *Elements of Modern X-ray Physics*. Wiley, Chichester, England.

[4] Bacon, G.E. (1955). *Neutron Diffraction*. Clarendon Press, Oxford.

[5] Baym, G. (1969). *Lectures on Quantum Mechanics*. Benjamin Publishing, New York.

[6] Bée, M.J.Y. (1988). *Quasielastic Neutron Scattering*. Adam Hilger, Bristol.

[7] Blakemore, J.S. (1985). *Solid State Physics*. Cambridge University Press.

[8] Bohr, N. (1913). On the constitution of atoms and molecules. *Philos. Mag.*, **26**, 1–24, 476–502 and 857–75.

[9] Bragg, W.H. and Bragg, W.L. (1913). The reflection of X-rays by crystals. *Proc. Roy. Soc.*, A**88**, 428–38.

[10] Bragg, W.L. (1912). The diffraction of short electromagnetic waves by a crystal. *Proc. Camb. Phil. Soc.*, **17**, 43–57.

[11] Brockhouse, B.N. (1955). A new tool for the study of energy levels in condensed systems. *Phys. Rev.*, **98**, 1171.

[12] Broglie, L. de (1923). Radiation – ondes et quanta. *Comptes rendus*, **177**, 507–10.

[13] Bubeck, D., Filman, D.J., Cheng, N., Steven, A.C., Hogle, J.M., Belnap, D.M. (2005). The structure of the poliovirus 135S cell entry intermediate at 10-angstrom resolution reveals the location of an externalized polypeptide that binds to membranes. *J. Virol.*, **79**, 7745–55.

[14] Byrne, J. (1994). *Neutrons, Nuclei and Matter*. Institute of Physics Publishing, London.

[15] Chakin, P.M. and Lubensky, T.C. (2000). *Principles of Condensed Matter Physics*. Cambridge University Press.

[16] Compton, A.H. (1923). A quantum theory of the scattering of X-rays by light atoms. *Phys. Rev.*, **21**, 483–502. The spectrum of scattered X-rays. *Phys. Rev.*, **22**, 409–13.

[17] Cooley, J.W and Tukey, J.W. (1965). An algorithm for the machine calculation of complex Fourier series. *Math. Comput.*, **19**, 297–301.

[18] Cooper, M.J., Mijnarends, P.E., Shiotani, N., Sakai, N., Bansil,

A. (2004). *X-ray Compton Scattering*, Clarendon Press, Oxford.

[19] Descartes, R. (1637). *Discours de la méthode pour bien conduire sa raison, et chercher la verité dans les sciences*. Translated by Laurence J. Lafleur (1960) as *Discourse on Method and Meditations*. The Liberal Arts Press, New York.

[20] Dirac, P.A.M. (1930). *The Principles of Quantum Mechanics*. Clarendon Press, Oxford.

[21] Dorner, B., Burkel, E., Illini, Th. and Peisl, J. (1987). First measurement of a phonon dispersion curve by inelastic X-ray scattering. *Z. Phys. B Condens. Matter*, **69**, 179–83.

[22] Eckert, J. (1992). Theoretical introduction to neutron scattering spectroscopy. *Spectrochim Acta*, **48**A, 271–83.

[23] Fermat, P. de (1662). Letter sent to *Cureau de la Chambre*, 1st January.

[24] Fermi, E. (1950). *Nuclear Physics*. University of Chicago Press.

[25] Fresnel, A. (1823). Mémoire sur la loi réflexion imprime a la lumière polarisée. *Mém. de l'Acad.*, **11**, 393–433.

[26] Friedrich, W., Knipping, P. and Laue, M.T.F. von (1913). Interferenzerscheinungen bei Röntgenstrahlen. *Ann. Phys., Lpz.*, **346**, 971–88. See also *Sitz. Math. Phys. Kl. Bayer Akad. Wiss.* (1912) 303–22.

[27] Guinier, A. (1939). La diffraction des rayons X aux très petits angles: Application à l'étude de phénomènes ultramicroscopiques. *Ann. Phys.*, **12**, 161–237.

[28] Guinier, A. and Fournét, G. (1955). *Small-Angle Scattering of X-rays*. Wiley, New York.

[29] Hannon, A.C., Howells, W.S., Soper, A.K. (1990). ATLAS: a suite of programs for the analysis of time-of-flight neutron diffraction data from liquid and amorphous samples. *I.O.P. Conf. Series*, **107**, 193–211.

[30] Hayter, J.B. and Penfold, J. (1984). Use of viscous shear alignment to study anisotropic micellar structure by small-angle neutron scattering. *J. Phys. Chem.*, **88**, 4589–93.

[31] Hendrickson, W.A. (1991). Determination of macromolecular structures from anomalous diffraction of synchrotron radiation. *Science*, **254**, 51–8.

[32] Hughes, V.A., Rees, P., Brown, M.R. and Roser, S.J. (2007). On the use of diffuse neutron reflectivity to probe domains in phospholipid layers. *Eur. Biophys. J.*, **36**(S1), S74.

[33] Huygens, C. (1678). *Traité de la Lumière*.

[34] Karle, J. and Hauptman, H. (1950). The phases and magnitudes of structure factors. *Acta. Cryst.*, **3**, 181–87.

[35] Kittel, C. (1953). *Introduction to Solid State Physics*. Wiley, New York.

[36] Koester, L. (1977). *Springer Tracts in Modern Physics*, Vol. **80**, Springer-Verlag, Berlin.

[37] Kwan, A., Dudley, J. and Lantz, E. (2002). Who really discovered Snell's law? *Physics World*, **15**, 64.

[38] Laplace, P.S. de (1812). *Théorie Analytique des Probabilités*. Cour-cier Imprimeur, Paris.

[39] Lekner, J. (1987). *Theory of Reflection*, Martinus Nijhoff Publishers, Dordrecht.

[40] Longair, M.S. (1981). *High Energy Astrophysics*. Cambridge University Press.

[41] Lovesey, S.W. (1986). *Theory of Neutron Scattering from Condensed Matter*, Volumes 1 and 2. Clarendon Press, Oxford.

[42] Lovesey, S.W. and Collins, S.P. (1996). *X-ray Scattering and Absorption by Magnetic Materials*, Clarendon Press, Oxford.

[43] Materlik, G., Sparks, C.J. and Fischer, K., eds. (1994). *Resonant Anomalous X-ray Scattering: theory and applications*. Elsevier.

[44] Mezei, F. (1972). Neutron spin echo: a new concept in polarised thermal-neutron techniques. *Z. Physik.*, **255**, 146–60.

[45] Mihas, P. (2005). Use of history in developing ideas of refraction, lenses and rainbow. Demokritus University, Thrace, Greece.

[46] Mitchell, P.C.H., Parker, S.F., Ramirez-Cuesta, A.J. and Tomkinson, J. (2005). *Vibrational Spectroscopy with Neutrons*. World Scientific, Singapore.

[47] Parker, S.F. and Haris, P.I. (2008). Inelastic neutron scattering spectroscopy of amino acids. *Spectroscopy*, **22**, 297–307.

[48] Parratt, L.G. (1954). Surface studies of solids by total reflection of X-rays. *Phys. Rev.*, **95**, 359–69.

[49] Patterson, A.L. (1934). A Fourier series method for the determination of the components of interatomic distances in crystals. *Phys. Rev. Lett.*, **46**, 372–6.

[50] Placzek, G. (1952). The scattering of neutrons by systems of heavy nuclei. *Phys. Rev.*, **86**, 377–88.

[51] Planck, M. von (1901). Ueber das gesetz der energieverteilung im normalspectrum. *Ann. Phys.*, **309**, 553–63.

[52] Porod, G. (1951). Die Röntgenkleinwinkelstreuung von dichtgepackten kolloiden systemen. *Kolloid Z.*, **124**, 83–114; **125** (1952) 51–7 and 108–22.

[53] Porod, G. (1982). *Small-Angle X-ray Scattering*, edited by O. Glatter and O. Kratky. Academic Press, London.

[54] Pynn, R. (1990). Neutron scattering: a primer. *Los Alamos Science*, **19**, 1–31.

[55] Rashed, R. (1990). A pioneer in anaclastics: Ibn Sahl on burning mirrors and lenses. *Isis*, **81**, 464–91.

[56] Rayleigh, J.W.S. (1912). On the propagation of waves through a stratified medium, with special reference to reflection. *Proc. Roy. Soc.*, **86**, 207–66.

[57] Reiter, G.F., Mayers, J., Platzman, P. (2002). Direct observation of tunneling in KDP using Neutron Compton Scattering. *Phys. Rev. Lett.*, **89**, 135505.

[58] Rutherford, E. (1911). The scattering of α and β particles by matter and the structure of the atom. *Philos. Mag.*, **21**, 669–88.

[59] Sears, V.F. (1989). *Neutron Optics*. Clarendon Press, Oxford.

[60] Sim, G.A. (1960). A note on the heavy-atom method. *Acta. Cryst.*, **13**, 511–2.

[61] Sivia, D.S. (1996). *Data Analysis: a Bayesian tutorial*. Clarendon Press, Oxford. Second edition (2006) with J. Skilling.

[62] Sivia, D.S. and David, W.I.F. (1994). A Bayesian approach to extracting structure-factor amplitudes from powder diffraction data. *Acta. Cryst.*, **A50**, 703–14.

[63] Sivia, D.S. and David, W.I.F. (2001). A Bayesian approach to phase extension. *J. Phys. Chem. Sol.*, **62**, 2119–27.

[64] Sivia, D.S. and Rawlings, S.G. (1999). *Foundations of Science Mathematics*. Oxford Chemistry Primers Series, **77**, Clarendon Press, Oxford.

[65] Soper, A.K. (2000). The radial distribution functions of water and ice from 220 to 673 K and at pressures up to 400 Mpa. *Chem. Phys.*, **258**, 121–37.

[66] Squires, G.L. (1978). *Introduction to the Theory of Thermal Neutron Scattering*. Cambridge University Press; reprinted by Dover, 1997.

[67] Sviridov, L.A., Battle, P.D., Grandjean, F., Long, G.J., Prior, T.J. (2010). Magnetic ordering in nitrides with the η-carbide structure, $(Ni, Co, Fe)_2 (Ga, Ge) Mo_3 N$. *Inorg. Chem.*, **49**, 1133–43.

[68] Van Hove, L.C.P. (1954). Correlations in space and time and Born approximation scattering in systems of interacting particles. *Phys. Rev.*, **95**, 249–62.

[69] Watson, J.D. and Crick, F.H.C. (1953). Molecular structure of nucleic acids. *Nature*, **171**, 737–8.

[70] Willis, B.T.M. and Carlile, C.J. (2009). *Experimental Neutron Scattering*. Clarendon Press, Oxford.

[71] Wilson, A.J.C. (1942). Determination of the absolute from relative X-ray intensity data. *Nature*, **150**, 151–2.

[72] Wilson, A.J.C. (1949). The probability distribution of X-ray intensities. *Acta. Cryst.*, **2**, 318–21.

[73] Windsor, C.G. (1981). *Pulsed Neutron Scattering*. Taylor and Francis, London.

[74] Wolf, K.B. (1995). Geometry and dynamics in refracting systems. *Eur. J. Phys.*, **16**, 14–20.

[75] Woolfson, M.M. (1956). An improvement on the 'heavy-atom' method of solving crystal structures. *Acta. Cryst.*, **9**, 804–10.

Index